Copyright © 2021 by **Jaco Friedrich**

English translation by **Maaike van Westen**

All rights reserved. This book or any portion thereof may not be reproduced or used in any manner whatsoever without the express written permission of the publisher except for the use of brief quotations in a book review.

These columns previously appeared in Dutch in Bits&Chips and Mechatronica&Machinebouw (Techwatch).

Coorporation and communication

How can I listen better?	1
The power of asking an open-ended question	3
How do I handle my feelings when I am criticized?	5
How do I give feedback without hurting the relationship?	7
How can I be critical without scaring people off?	9
How can I increase mutual trust?	11
How do I steer the conversation in a positive direction?	13
How can I be assertive without being rude?	14
How do I handle jealous coworkers?	17
How do I support a coworker who is being bullied?	19
How can I handle cultural differences?	21
How do I become part of the team?	23
What should I say at the coffee machine?	25

Leadership

Stakeholder management, what does it mean?	28
How do I get people to really listen when I present a solution?	30
As a specialist, How can I take on more of a leadership role in the organization?	32
How do I make decisions based on incomplete information?	34
What does the board focus on when I present a solution?	36
What should I do when someone attacks me?	39
How can I quickly align with my key stakeholders?	41
How do I deal with the vague suspicion that I'm being deceived?	43
How do I get the organization back in line?	45
How do I deal with ideas that are treated as "sacred"?	47
What should I do when no one wants to listen?	49

Team management

What's the best way to delegate work?	52
How do I become a good mentor?	54
How can we become a strong project team?	56
How do I get a quiet group to talk?	58
How can I be less bossy and still be effective?	60
How do I motivate my team to develop and perform better?	62
How do I talk to someone about their annoying behavior during meetings?	64
What is the best approach to mentoring junior coworkers?	66
How do I tell someone the bad news?	68
How do I recognize the three signs of a looming burnout?	70
How do I develop a training portfolio that meets our needs?	72
How can I handle the organization's rapid growth?	74
How do my team and I come to terms with the death of our colleague?	76

Consultative sales

How can I get better at selling my service?	79
How do I call a potential customer?	81
How do I persuade potential customers to meet with me, even if I don't know them?	83
How do I put pressure on a negotiation without shutting the door?	85

Self-management

How do I make sure I'm doing the important things at work?	88
How can I remain effective in chaotic circumstances?	90
How do I become more patient?	92

How do I come across as more senior?	94
How do I develop more empathy?	96
How do I keep developing without changing jobs?	98
How do I retain the skills I have learned?	100
How do I show leadership and take a step towards a senior position?	102
How do I apologize?	104
More information about Jaco and his writings & training	106

Coorporation and communication

How can I listen better?

A software engineer asks:

During a recent performance review, my manager told me that coworkers think I don't listen carefully enough. He pointed out that during our meeting, I didn't always follow what he was saying, I changed topics, and I interrupted him at the wrong moment. How can I learn to listen to my conversation partners?

The communication trainer answers:

Technical people get paid to solve problems. But they often work on complicated projects with a lot of different colleagues. To be able to do that, you have to collaborate, and that means you have to communicate. You need a relationship to do so. Think of it as a telephone line: If you have a bad connection, you can't convey your message.

So you need a connection, or there won't be any communication. Making a connection means: focusing on your conversation partner with your body and your mind. Don't start checking your email and listen with half an ear. Just give the other person your full attention. Turn your body towards your conversation partner. Look at him or her periodically, but avoid staring since that can come across as intimidating. Focus your attention on the other person. You can mirror the body posture of your conversation partner, meaning that you can sit in more or less the same way. That gives a sense that "we're in this together." Just don't exaggerate, because then it seems fake and that's annoying.

Speak at about the same speed as the other person. This is called "pacing." People who speak very fast will need to slow down, so they don't outpace the person they're talking to. Similarly, someone who has a tendency to speak slowly will need to pick up the pace to keep engaging a colleague who speaks faster. If you mirror someone and pace your speech, you greatly improve your chances of having the conversation go smoothly.

In terms of content, how do you make sure you really listen and understand each other? You do this by using active, or comprehensive, listening techniques: listen, summarize, and ask follow-up questions (LSA).

Start by listening, just listening. Don't think or try to come up with your next question, just pay attention. Forget about your own agenda for a moment, it will still be there when you get back to it. Simply pay attention to what the other person is saying, and in the meanwhile your brain will process the incoming information without any conscious effort on your part. The right question or the feeling that you want to say something will come up automatically. So relax and stay focused.

Next, it's important to periodically summarize what you've heard. For example, you could say: "OK, so you're saying X,Y, and Z. Is that right?" This way you can make sure you understand the other person correctly. That's helpful for them, because they can tell you're really listening, and they can clear up any misunderstandings. This will give them confidence in you as a listener.

Summarizing increases trust in you as a listener

Summarizing gives the listener another advantage: It provides a moment of calm. The speaker takes a pause while you are summarizing. They will need to think it over and will either agree or elaborate on their story. That pause is a good moment to ask just the right question, for example: "You just said A and B, which was not completely clear to me. Could you please explain?" As a listener, you create a structure by doing this and you keep leading the conversation.

Do you want to interrupt the other person? Just do it. There will never be a great moment to do this, so just go ahead. This is how it works: Be decisive, speak a bit louder, lean in with your body, and start by saying something positive. For example: "Interesting, what I wanted to say is... and..." Are you someone who always interrupts others? Stop doing that and start listening patiently.

The power of asking an open-ended question

A senior project leader asks:

I started out as an engineer and worked my way up to lead engineer and project leader, because I tend to speak my mind and get things done. Now I'm starting to notice my own limitations. It seems my team is always looking to me for a solution. If someone brings up an idea, team members usually check my reaction before they respond. I think this gives me a more dominant position than is truly useful. It is limiting my team's development and my own. How do I get the team to provide more input and take more responsibility?

The communication trainer answers:

"What Got You Here Won't Get You There" is the title of a book by Marshall Goldsmith. This sums up the situation of a senior project leader who is limiting himself and his team in their development. The very qualities that helped you reach your current level can block further development.

Let's look at this example of a project leader. He has grown into his role of senior project leader because he was capable of tackling the work and achieving results. Now he is discovering his own limitations. Since he has always had an active role, his team is used to him providing solutions or coming up with better ideas when a team member presents a proposal. Maybe his ideas are generally better. Nevertheless, it's important for him to take a step back.

Team members are apprehensive about advocating for their own ideas because they will inevitably be overruled. After all, the rest of the team will side with the project leader. This makes it unsafe to speak your mind. People become reactive. In this situation, new team members are also less likely to be successful than they would be otherwise. The same is true for the senior project leader's potential successors. After giving it a try, they will leave.

How do you prevent this team culture from taking hold, and what steps do you take if it has already taken hold in your team? The most important step is to act less dominant as a leader.

Here's how you do it: Let's say there is a problem that needs to be solved. Normally you might describe the problem, propose a solution, see what people think, and tell someone to implement the solution. Instead, bring up the problem and ask an open-ended question: "What is the best way to solve this problem?" Make sure your words and your body language are in line with each other. Faking it won't work. So show people non-verbally that you're genuinely curious about what they have to say. You do this by briefly leaning back, relaxing, looking around calmly for a moment (without staring), and pausing.

"What got you here, won't get you there"

As a quick aside: Many indigenous tribes in North America used a so-called talking stick for their meetings. The person holding the stick had the right to speak. This prevented people from interrupting each other, it was clear whose turn it was, and the speaker could give someone else a turn. Now back to the project leader.

After asking your team for their thoughts, you put down the talking stick for someone to pick up, as it were. The ball is now in their court. Is no one speaking up? Repeat your question and say you would really like to hear all their ideas. Relax again. Save your own idea for a solution for last, if you need it at all. It's best to choose a team member's idea that works, even if you have an idea that might work slightly better.

Team members will become increasingly independent when you give them more responsibility. Do this consistently. Sharing the ball allows you to enlarge your own scope.

How do I handle my feelings when I am criticized?

An engineer asks:

People think I don't respond well to situations where I don't like the solution that was chosen. I don't deal well with criticism either. I will start defending myself or attack the other person for their opinion. This creates quite a bit of turmoil. If I could get better at accepting that things don't always go my way, it would probably benefit me in the end, and I can learn from criticism too. How should I handle this?

The communication trainer answers:

First, let's see what happens when we don't listen or "don't respond well." Our heads are generally filled with thoughts 24/7. No one really knows where they come from, or where they go. The fact is, we're pretty busy with our thoughts, and even when we don't know exactly what they are, we take them very seriously. Seriously enough to act on them. Add our emotions to the mix, and things can quickly heat up.

Research shows that in a fit of anger, only 10% of our perception is still based on facts. This can easily happen when we're faced with something we don't like or when we're criticized. Our thoughts and emotions run off with us, so to speak. Taking them too seriously is a recipe for disaster. So what's the alternative?

Listening to something you don't want to hear or that hurts your feelings is not easy. You will need to ignore your hardwired response. This is generally one of the following: attack ("yes, but your idea doesn't make sense because A or B…"), defend ("I think I'm right because…"), or lose your train of thought ("uh…"). A response like this takes only a split-second and barely involves the thinking brain. It means you'll have to pay attention and bite your tongue. There's no other way. Put aside your hurt ego for a moment, and listen.

How do you do this in real life? Say your boss is criticizing you ("The meeting failed because your idea made no sense"), and you feel a reaction coming up. Now do the following:

1) Take a deep breath (you can even literally bite your tongue; the pain will pull you from your emotional state into your sensory perception).

2) Lean back slightly, relax, and calmly look at the other person.

3) Say "Please tell me, how do you view this?". Be interested and try to understand. Now the other person will explain their point of view. This will buy you some time to recover and start thinking "normally" again.

4) Ask follow-up questions until you understand exactly what the other person means. This should reveal if they have a point or not. Maybe it was more bark than bite, and that should be clear too by now. Also examine your own assumptions and frame of reference. Maybe someone says something that sounds impossible to you. If you listen carefully, it may not seem so impossible after all.

5) Summarize the other person's point of view ("Okay, so you think that X and Y..."). They now know you understand them, which has a relaxing effect. The discussion will be slowed down, which is another thing that defuses tension.

UNDERSTANDING AND AGREEING ARE NOT THE SAME THING.

---

6) At this point you finally express your own reaction. You should realize that listening is completely separate from whether or not you agree. Understanding and agreeing are not the same thing. Find out what someone is trying to say regardless of your personal feelings. If you can do that in a relaxed way, you will be more successful at making your own point, should that still be relevant. Any hurt feelings are simply part of having a (work) relationship. Just remember: The cut is only skin deep.

How do I give feedback without hurting the relationship?

An engineer asks:

As a senior engineer, I supervise the work of other engineers, but I'm not always successful at it. For example, I think one of my engineers should do things differently, but he doesn't seem to hear my criticism. I'm worried that I'll have to redo the work myself.

A similar question: How do I get the engineer to listen to my criticism?

I need information from one of my coworkers. I have already asked him a number of times but am not making any progress. It annoys me. Because I'm so annoyed, I worry that I will sound snippy when I say something about it. How should I handle this?

The communication trainer answers:

Both of these situations involve providing feedback. That is, you need to tell someone what you think about something in order to improve the situation. It's a difficult soft skill to master, especially when it concerns negative criticism. There are two pitfalls: You either tiptoe around the problem and don't convey your message, or you are too rude and damage the relationship. We will often try to avoid these pitfalls by not saying anything. It's obvious that this will not make the problem go away. On the contrary, the problem will get worse. When remain silent for a long time and finally speak up, chances are that all that bottled-up criticism will come out the wrong way. You end up exactly with the kind of pointless discussion that you were trying to avoid in the first place.

If you're not careful, you might even draw the conclusion that it's a bad idea to speak up the next time around. This way, the threshold for providing feedback gets higher. And that's bad news, because as people we can only grow by receiving feedback. If we're not aware of what we're doing and what the effects are, we cannot adapt our behavior to what is required. In short, if we want to develop, we will need feedback from others, whether we like it or not. The same is true for your colleague. It's easier to say something when you keep this in mind. After all, you speak up to improve the situation and to help the other person grow.

Follow these four steps to provide effective feedback, to influence your colleague's behavior. You want to make sure your feedback is heard and you don't put the relationship under pressure. All of these steps are necessary, so follow them one by one.

1) Mention that you want to say something about the work or collaboration (so don't address the actual feedback yet). Now the other person knows they have to pay attention. For example, you can say: "I would like to talk to you about something I've noticed (or that has been bothering me), is that OK?" You basically agree on taking a moment to talk to each other. Make sure that your personal motivation is to improve the situation, not to put someone in their place. This will help to create an open conversation.

2) Be specific and factual in addressing the other person's behavior and the effect it has on you and/or on the work (this could even include an emotion). Don't say: "The way you're handling this is wrong." That's unclear. For example, you can say: "I saw you handed work in later than agreed for the third time (the other person's behavior). That means I get behind with my work, and I don't have enough time to do a good job (how it effects the work). I'm concerned about getting stuck, and this worries me. It also annoys me that you don't respect our agreement (how it effects me). This is important, because we often forget to mention the effect something has on us. And if we don't mention it, our message often misses the mark.

3) Take a step back and let the other person respond. Do this by asking a question. For example, you can say: "Do you recognize this?" or "What do you think about that?" Wait for the answer (a pause of at least four seconds will prompt someone to respond). This can make them a bit uncomfortable. If that's the case, your message has gotten through. It's important to just let that sit. Don't try to move on to a solution too quickly. Allow the uncomfortable feeling to transform into motivation: "How can I improve?" And that's exactly what you need to take the next step.

4) Try to find a solution together.

Being good at providing feedback is no guarantee for success. However, it does give you the tools to positively influence most situations. Best of luck!

How can I be critical without scaring people off?

A systems architect asks:

As a systems architect, I often review my colleagues' work and I mentor junior employees. They feel I'm too aggressive when I ask questions. Even though that doesn't really bother me, I've noticed others are very careful around me and are starting to avoid me. That's not my intention. How can I be critical without scaring people off?

The communication trainer answers:

Work will not improve without scrutiny. The occasional figurative punch can shake things up. After all, if someone keeps making the same mistake over and over, a careful comment is not going to help. Confrontation can act like a wake-up call. However, some people make a habit of confronting their colleagues. This approach certainly has its limits. In order to learn, one needs the courage to open up. You gain very little when a colleague gets defensive, withdraws, or stops talking because of your comments or critical questions. You probably don't mean to do this, but sometimes it happens before you even realize.

The resistance people feel is largely caused by the way you ask questions. How does that work?

There are basically two types of questions: open-ended questions and closed-ended questions. An open-ended question asks someone to provide information and makes them think. Open-ended questions start with what, who, which, where, why, and how. A close-ended question can be answered with "yes" or "no." For example: "Do we have an agreement?" "Is everything clear?" It forces the recipient to give a very specific response.

An open-ended question is an invitation to be transparent. For example: "What made you decide to do it this way?," "Who would be the best person to ask?," "How did you reach this conclusion?," "Which approach do you think works best?," or "Where should this be implemented?" Questions starting with Why are also on the list, but this gets tricky. A question like "Why did you do it this way?" will often come across as an attack. That's no problem if you're talking to a friend or co-worker you know well. However, if you use why in a tense situations or with someone who is lower on the corporate ladder, you can expect a defensive response. The goal of your open-ended question is to create an open dialog with your colleague, but instead they will build up resistance and stop listening.

If you have an opinion, express it. If you have a question, ask it.

You will also be met with resistance if you phrase your opinion as a question. For example: "Don't you agree this never really works?" The sentence may end in a question mark, but that doesn't make it an effective question. The recipient can tell you have an opinion, feels manipulated, and will go on the defense. To prevent this from happening, follow this rule of thumb: "If you have an opinion, express it. If you have a question, ask it." For example: "I don't think this will work, because of X, Y, and Z," followed by "What do you think?" Because you express your opinion, it's clear where you stand. And by asking "What do you think," you put the ball back in their court.

Reinforce this with your body language. Here's how it works. Pose an open-ended question. You can start by stating your own opinion, e.g. "I think A and B... What do you think?" Relax, lean back a bit, and pause without staring (pause for at least four seconds, so the brain has time to process your question). You are transparent and provide space for your colleague to answer. This way, you create an open work environment that allows for an easy exchange of information. There will always be plenty of opportunities to put more pressure on someone.

How can I increase mutual trust?

A team manager asks:

As a group leader, I often work with various project leaders. I make sure that projects are staffed and that we deliver the right product. However, I made an unfortunate mistake in the project budget and was ashamed to admit it, so I tried to hide it. I've recently noticed that not everyone believes me when I make statements about the expected amount of work, planning, and other matters. That bothers me. It feels as if they don't trust me. How do I improve the situation?

The communication trainer answers:

It's easier to lose someone's trust than to gain it. That means you will need to work hard to gain people's trust, and it only takes a few bad mistakes to lose it. That would be a shame, because every aspect of teamwork is easier when you trust each other.

But how do you build mutual trust? Can you force it? What's the alternative? A good starting point is to trust people exactly once. This way you don't approach every situation with mistrust, but you also don't let people walk all over you. If it turns out someone is deceiving you, you're done.

But should you simply trust someone for their big blue eyes until they prove the opposite? That would not be wise. It's good to use common sense before you put your trust in someone. What should you pay attention to? There's a nice formula for this. You can describe trust as the sum of transparency, capabilities, and reliability, divided by self interest.

Transparency is the extent to which you and the other person are open about your motives and actions. If you hide something and are exposed, others will immediately trust you less. It raises questions: "Why didn't he tell us? Is he hiding something?" Even when it concerns something very minor, on some level people will wonder if you're also hiding other, possibly more important things. It's better to prevent that from happening.

Aside from transparency, your actual capabilities are important too. In other words, what are your tangible achievements? This is also connected to reliability Do you know how to deliver the same quality time and time again? Are you doing what you've agreed to do? If that's the case, people's trust in you will grow significantly. In short: Do what you say and say what you do. Divide the sum of all of this by the degree of self interest you're showing. This last aspect is very important.

How does this work?

You could say that people have the innate tendency to share. That's good for the survival of our species, and it makes us happy. You can see this in little children. However, some people lose this tendency, and it doesn't become a part of their adult behavior. When they have to collaborate, this results in friction. We simply trust someone more when they think of us too, and not just of themselves.

We simply trust someone more when they think of us too, and not just of themselves.

If your actions are solely driven by "me, myself, and I," they will be less effective. It may work in the short term. Say you need a bike, and you decide to steal one. Even though your actions may pay off in the short term, it is based on a limited view of a situation. What if it turns out that the bike you've stolen belongs to the neighbor you share a garden with? It's more realistic to broaden your motivations by including more people—you will automatically gain more support and will therefore be more successful. Do this by keeping the interests of others in mind when you make plans.

Have you lost people's trust and would you like to gain it back? You will have to do better in all aspects. The truth is, actions speak louder than words, so you will have to show real improvement in your behavior.

How do I steer the conversation in a positive direction?

An engineer asks:

When I am talking to one of my coworkers, conversations often don't go smoothly. Later on I'll realize I was drawn into a discussion that wasn't very productive. I would like to avoid having these discussions and steer the conversation in a positive direction. How can I do that?

The communication trainer answers:

A conversation between two people is a dynamic process with different elements at play, including substantive interests, personal preferences, and emotions. That's why it doesn't take much for a conversation to become difficult or unproductive. How do you recognize what is happening, what position you are in, and what exactly the other person is doing? And what can you do to make a conversation productive again?

Timothy Leary, an American psychologist from the last century, developed a model that provides insight into interactions and can serve as an aid for changing an unproductive conversation into a productive one. According to Leary, the dynamics of any interaction can be boiled down to two features that are always present: intimacy and authority. In other words, are we in opposition or are we collaborating, and who is the boss? This creates two axes: "against versus together" and "above versus below."

If you are above and against, you're acting bossy by giving orders and challenging people. This can be functional, for example when a fire breaks out or when there's no time to discuss a topic. However, it will often be met by resistance. In the "below-against" position, you will either give someone pushback or withdraw. "Above-together" means you are positioning yourself as a cooperative leader: You take the initiative, but you also pay attention to the interests of others. You realize that you cannot do it alone and that you need others. In the "below-together" position, you will be helpful and willing to work on the ideas that are presented to you. Needless to say, working in the "together" zone is more productive in a lot of situations and generally experienced as more enjoyable.

The dynamic between different positions involves two important mechanisms: above triggers below and vice versa (complementarity), against triggers against, and together triggers together (symmetry). These mechanisms can work against you, but you can also use them to your advantage if you pay attention. If someone acts in a dominant way (above-against), it can automatically trigger below-against behavior. The reverse is true too: If someone becomes uncooperative or obstructionist, the other person will have a tendency to apply more pressure. But if someone positions themselves as "together," you will be more inclined to adopt a cooperative attitude too.

If you don't pay any attention to this and respond impulsively, you can quickly get caught in a pattern that may hamper successful collaboration. Instead, you will need to closely observe what is happening in the moment and respond in a "counterintuitive" way, as it were. You do this by using the same mechanisms: complementarity and symmetry.

Closely observe what is happening and respond in a "counterintuitive" way

---

When someone positions themselves as above-against, don't oppose them. Instead, try to respond from a position of together-above. For example, explain the situation clearly, and if necessary give feedback about what you see happening with the other person or about any (negative) effects this has on you or the work. That way, you take the lead in a positive manner, stay respectful, and invite the other person to change their behavior and cooperate. Are you dealing with someone who is passive or obstructionist? Instead of applying pressure, it may be more helpful to take on a below-together attitude by calmly asking an open-ended question and showing that you're interested, thereby inviting them to take on a more active role. This will require some practice, but even just understanding what is happening will already help you improve the interaction.

How can I be assertive without being rude?

An engineer asks:

My boss tells me that I need to be more assertive. I agree, but when I see people acting assertive, it often puts me off. They're loud and pushy. That just not me, and I don't want to be that way. How can I become more assertive in a way that suits me?

The communication trainer answers:

What is "assertive" or "aggressive"? Of course this all depends on the definition. But when you think, "Oh boy, that is one assertive person," chances are they are aggressive rather than assertive. What's up with that? Let's take a look at assertiveness.

Assertiveness means that you respect yourself and the other person. You advocate for your own interests but also pay attention to someone else's interests. Within this model of assertiveness, it would be considered aggressive when you only act out of your own interests without any consideration for others. You can score a win, but the relationship will be damaged.

Let's say a coworker asks you to do something. It's not the first time you've stayed late to help this colleague. He has trouble at home, and you understand this is hard for him. This time, you don't feel like helping out, and you say: "No, I'm not doing it today! Don't you understand the meaning of the word No?" Your coworker will certainly understand that you don't want to do it. It may even give you a momentary sense of relief. But it will put stress on the relationship.

If you're all the way on the other end of the assertiveness spectrum, you will be more considerate of the interests of others. In that case, you are easily taken advantage of. This would be called "sub-assertive." In the above example with a coworker, it would sound like this: "Okay, I understand... you need help. Um... I'll take care of it." You give in, even though it bothers you. The relationship may seem fine, for now, but you haven't made your point. You start bottling up your emotions.

Every time you allow someone to overstep your boundaries, even when it concerns a minor issue, you build up tension and become insecure or stressed. That bottle of emotions will be full at some point. A relatively minor event may trigger you to react to the entire series of events. You explode and may become aggressive. Maybe the person you're targeting wasn't even involved in all the events leading up to this point, but she's in the line of fire and she is in shock.

You might draw the wrong conclusion: "See, it's better to keep my mouth shut." A better conclusion is that you should have addressed the issue with the right person sooner, before it became so emotionally charged. That's being assertive. You pay attention to the relationship (that is, the importance of other people's feelings), and to your own needs. You work on both aspects simultaneously. Assertiveness is key to long-term cooperation.

Practically speaking, this is how you make your point:

1) Use the word "I," for example "I think it's important that..."
2) Keep your message short and clear.
3) Use one or two arguments to support your point.
4) Repeat your message if necessary.

At the same time, work on the relationship by doing the following:

1) Ask questions: "What is going on exactly?"
2) Ask follow-up questions until your picture of the situation is complete.
3) Show understanding, for example: "Oof... yes, that sounds really difficult."
4) You can also summarize the situation, e.g. "So if I understand correctly, you're saying..."

If you speak honestly about both of these aspects, you can clarify your position and at the same time work on the relationship. In this example, it could sound like this: "What's going on? ...So you're stuck? Too bad you're dealing with this again. It's a difficult situation... I do notice that I've covered for you three times already this week. I'm just tired. I would really like to go home on time. That's why I'm saying "no" this time."

> **It's crucial to find out what's going on with the other person and be honest about what you want.**

Needless to say, there are situations that cannot be solved by assertiveness alone. If someone wants to steal your bag and you can take them on, it may be helpful to yell: "That's my bag, get your hands off!" But if the thief looks dangerous, you're better off acting sub-assertive by making yourself invisible for a moment. In most situations, however, you will want to show respect for yourself and the person you're working with, and weigh their interests as well as your own when making choices. Assertiveness is key to long-term cooperation that works for both parties, even at times when you disagree.

How do I handle jealous coworkers?

A team manager asks:

I was recently promoted to team leader. It's going well, and I received a lot of positive feedback about some improvements I recently implemented. To my surprise, I have also noticed that some people are trying to undermine me in an awful way, which is not something I would have expected from them. What exactly is happening, and how do I handle this?

The communication trainer answers:

When you're doing well, it can evoke all kinds of feelings in others. Jealousy is among the worst of these. Sometimes people simply cannot stand it if others have more money, status, or success. It would be better if neither of you had anything.

Jealousy can drive people to breach the boundaries of decency. It is a many-headed monster that can manifest itself in different ways. For example, people may speak badly of someone behind their back, withhold information, refrain from doing something just to trip someone up, or purposely let someone take on an impossible task so they're bound to fail.

It will often be difficult to pinpoint who is behind it. You might notice that you are experiencing resistance for no valid, conceivable reason. Or you may be falsely accused. Nothing human is alien to us.

Sometimes, you may even do this yourself without realizing it. Say your coworker is making enormous progress and is praised by everyone. Next time she presents a good proposal; you may find that you're less enthusiastic. You may even be overly critical. Maybe you don't really mean to sabotage this person, but you wouldn't mind if he or she stumbled a bit.

Although this undermining behavior is often unconscious, it can also be intentional. Spotting someone's underlying motives can be difficult when they use arguments that sound reasonable. Anyhow, they are clearly no longer discussing the matter at hand or acting in the company's best interest.

What is a smart way to handle this situation? It's important to recognize it if someone is undermining or badmouthing you. This often involves portraying you or your project in a way that is not based on facts, but on half truths and fiction. You counter this by putting out the whole story or by challenging the person who is undermining you to back up their statements. Do this sooner rather than later, because when a lie is repeated often enough, it becomes the truth, at least in the minds of many. Squash it quickly if you notice this is happening.

By not taking success personally, you avoid floating up like a balloon

---

It's also smart to behave in a way that prevents jealousy. You do this by not beating your chest when you achieve some success and by always staying modest. This doesn't mean you should let people take advantage—it just means that you don't brag about your success. Besides, you didn't achieve this success alone but with your team and all the other people who helped you. By not taking success personally, you avoid floating up like a balloon and becoming an easy target. Keep your feet on the ground, and know where you're coming from.

Do you get jealous too sometimes? Instead of feeling resentful, try to enjoy the other person's success. You will notice good things happen when you do. Success multiplies and lights you up too. Everybody happy.

How do I support a coworker who is being bullied?

An engineer asks:

I've noticed that a coworker on our team is often being bullied. It really bothers me, but I'm not sure what to do about it. I dread the thought of becoming a target myself. What can I do without getting too involved?

The communication trainer answers:

Bullying is systemic, transgressive, undermining behavior, often targeting one or more individuals. So it's not just one nasty comment or ignoring someone once. It's part of a pattern. And that's what makes it so painful.

If it happens only once, it could just be an incident. The second time should ring an alarm bell. Three times means there's a pattern that will continue until something is done about it. So if you want to put an end to the behavior early on, you will need to step in.

Bullying in the workplace can be many things. How do you recognize bullying or harassment? It can be a desire to humiliate someone through hurtful comments, name calling, or sexual intimidation. The perpetrator may be downright abusive or be more subtle and say things that just cross the line. A bully may regularly withhold information to make someone look bad, or exclude or ignore someone.

The person who is targeted will often be a bit different from the rest of the group. Maybe it's the way they talk, where they're from, the way they dress, or the fact that their position is temporary and they aren't part of the regular team. Even coming across as slightly more vulnerable than others can be a reason. Basically, it happens because it can happen.

How does being bullied affect someone mentally? Brain researchers had a group of volunteers stand in a circle and throw each other a ball. One of the participants was unaware that the others had been instructed not to play the ball to him. The researchers found that the areas of the brains that lit up, were the same as those for physical pain. So it literally hurts, and this all takes place within minutes.

People usually respond by trying to man up and act like they don't care. That usually doesn't work. The employee puts up a wall and is afraid to show who they are for fear of being ridiculed or bullied.

People who are being bullied, will often start to believe they are at fault. Their self-confidence is damaged even further by feelings of shame about being targeted and guilt because after all, shouldn't they be able to cope? This gives bullies even more license to keep going. They are not experiencing any boundaries.

Break the taboo and talk about it.

Does this all happen without anyone noticing? Of course not. There are always witnesses. Most of them would rather stay quiet, or they may even decide to take part in the bullying. They may be afraid to get dragged into the situation and become a victim themselves. Something similar happens when someone falls in the water and ten people are standing by without doing anything. They are all thinking the same thing: "Surely someone else will jump in." When there is only one witness and you ask someone for help personally, there's a better chance they will help you.

What can you do if you're a victim of bullying? You can become more assertive with the help of coaching or training. Weight training or karate, for example, will increase your confidence and make it easier to stand up for yourself. It's also recommended to get help. Find someone you trust and discuss how the situation can be improved. Can't find anyone? Consider if it's worth looking for a different job. It can be useful to find a career adviser who can help you with that.

What can you do if you witness bullying? Break the taboo and talk to the victim. Work together to find support. Companies can prevent bullying by creating a culture in which employees are held accountable for their behavior, managers are available for regular meetings, and harassment and bullying can be reported confidentially. Introduce a zero tolerance policy. Bullying is bad for business.

How can I handle cultural differences?

A team leader asks:

For one of our upcoming projects, we will be working with a foreign partner. I will be checking in with one of their staff members on a weekly basis. This is not something I look forward to, since he is from a different culture. In my experience, cultural differences often lead to miscommunication, which in turn can result in unwanted project delays. How can I avoid these kinds of communication problems?

The communication trainer answers:

Working with a foreign partner's employees can add an extra dimension to collaborating. You may be wondering: "Why are things not running smoothly?" Meanwhile, they are thinking: "The Dutch are a funny bunch." Sooner or later you will run into trouble if you don't account for differences. You will simply be caught off guard.

It would go too far to provide a primer for every culture here. However, there are a number of important elements you can pay attention to, according to Dutch social psychologist Geert Hofstede. He says you can describe a culture based on five dimensions.

The first one is power distance. Power differences in the Netherlands are quite small compared to other countries. This is evident, for example, in our almost instinctive aversion to authority. We make a strong case for this: After all, shouldn't content be our primary focus? In practice, this results in a very egalitarian way of organizing and in people being openly critical of their superiors (uh-oh, bad word). We actually expect everyone to ask critical questions about everything, starting in elementary school. In other words: Think for yourself and take responsibility. So if a manager proposes an idea, we will automatically question whether this is the best idea. The advantage is that the idea will often improve, but the disadvantage is that it can slow things down.

In short: Dutch employees tend to be critical and have a lot of freedom to make their own decisions. The opposite is true for many cultures around us. An employee in China will always ask himself what his boss might think and will let him make the final call. Without a manager's approval, any agreement between you and the employee will have little significance. Make sure you involve the manager.

The Dutch are a funny bunch

The second dimension is individualism. To what extent do you experience yourself as an individual or as part of a community? The Dutch are very individualistic. Being different is seen as more normal than in many of our surrounding countries. Individualism is reflected in the workplace, for example in the extent to which someone is willing to be vulnerable in front of their colleagues. Will you speak up, without feeling embarrassed, to say you don't know something or you disagree with the majority? In some cultures, doing so is not usual. Ask open questions when talking to colleagues from Japan (e.g. "How are you going to handle this?"). Wait to hear what they say and, more importantly, what is left unsaid. You will not hear a clear "No" very often. From their perspective, that would put you in an awkward and difficult position.

Masculinity/femininity is the third dimension. The more masculine a culture, the more competitive it is and the more important it is to be the very best at your job. In the Netherlands, the emphasis is on "working to live" and not the other way around.

The fourth dimension is uncertainty avoidance. This refers to the extent to which people accept uncertainty and handle things that don't go as planned. A co-worker from India will not be alarmed by a change. For someone from Germany, that may be different.

Finally, the fifth dimension: long-term orientation. This concerns the extent to which you prioritize short-term profits or a company's long-term continuation. Americans will often set short-term goals and are more likely to leave long-term effects out of the equation. In Japan, continuation in the long term is an important consideration. This can result in a somewhat fatalistic view.

So many people, so many faces. It pays to learn more about the culture of the people you do business with.

How do I become part of the team?

An engineer asks:

I have a new job at a new company, and have noticed that the team doesn't fully accept me. Certain meetings take place without me, even though it would make sense for me to be there because of my role, and I miss out on interesting projects. This bothers me, and I wonder what I can do about it.

The communication trainer answers:

When you're a new team member, you don't automatically become part of the team. A team is a dynamic system of relationships between team members, with its own culture. These are the unwritten rules on how to do things and how to interact. When you're new, you don't know the rules yet, and it will take some time, maybe even a long time, to learn them. Depending on how well the team culture matches your own personality, it may be easy or you may need to work harder to fit in.

Another important factor is how willing the existing team is to welcome new team members. If that doesn't happen or others have little tolerance for what they view as "strange" behavior, you risk staying an "outsider." Without intervention, this can result in an impossible situation and early departure. This is often unnecessary and certainly unwelcome.

An outsider is like a "carrier" for opinions and behaviors that the team considers as "outside the norm." This is exactly why it could make for an interesting addition to the team. Does the team emphasize cordial relationships and avoid harsh criticism? In that case, someone who does the opposite will soon be considered difficult. If the team has a strong sense of unity, someone who is very competitive may not be always be a good match. If substance and doing the right thing for the organization is a core value and the new team member is focused on rapid carreer development, it will be difficult for him or her to earn credibility.

Is protecting a team culture always a bad thing? No. Maintaining a certain culture fosters the strong sense of unity and focus that can make a team extremely productive. At the same time, there is a risk of stagnation and one-dimensional development, which limits innovation and openness. It's important for an organization to be alert to the non-negotiable core values that really matter and to areas that need improvement.

The team leader is responsible for upholding the team's core values while keeping any negative effects in check. That's easier said than done, since the team leader is part of the organization's system too. The challenge for a leader who comes from "outside," is to look at the team and find out what is needed. If you notice that the team doesn't show openness, it's important to always support a new team member and discuss the situation if necessary.

An outsider can be an interesting addition.

If you are an "outsider" and want to find or claim your spot, the first thing you need to do is figure out what makes you different. If the team emphasizes technical credibility (which is true for most technical companies, fortunately), try to work on a job or create a situation that allows you to show what you're worth. This doesn't always have to be a project—you can also lead the team with your unexpected talent for survival and ingenuity during a survivor team building game, for example. Ask a trusted coworker if your behavior differs from "normal," and then think about how you can adapt without losing your individuality.

Just realize that you and the organization both have a problem, and you'll have to work it out together.

What should I say at the coffee machine?

An electronics developer asks:

During my last performance review, my manager said that the quality of my work is good, but that I need to work on my communication skills. It's true that I'm often quiet at the coffee machine and during office outings, because I don't like chatting about trivial things. But my boss says small talk is necessary for my career. How do I overcome my reluctance and make casual conversation?

The communication trainer answers:

Small talk helps you connect to others. The ability to start a conversation with someone is an important skill. It allows you to easily get to know new people and "break the ice."

As a professional, you will need to focus your workplace communication on both content and the relationship with the other person. You are probably fine discussing content, which is a good thing, since you are paid to solve problems. Nonetheless, to be effective at exchanging information, you will also need to build a relationship with someone. The content of your message will get lost if your "connection" is full of static and interference.

In other words, you will need at least some positive contact to be able to discuss matters at all. And to make contact, you will need to say something. So what do you say at the coffee machine or during an office outing if you don't have anything substantive to talk about? Don't worry, this is something you can learn. The three most important things are your mindset, your physical posture, and what you say.

Establishing contact starts with your mindset: curiosity. Maybe you think: "I'm not particularly interested in this person." That may be, but this will make it harder to connect. Engage with curiosity. Start thinking about some questions, such as: I wonder who that is? What could he be working on? What is her specialization? If you take the initiative by saying something, curiosity will help you open up to what your conversation partner says.

As for your posture, don't be pushy, but also don't make yourself invisible. Relax.

Next up: What should you say? Maybe you have a question or observation. To draw someone out, always pick something that is right here, right now. For example, you can say: "Boy, that coffee is hot" or "Were you stuck in traffic too?"

Once you have caught someone's attention, you can play the ball back. Do this by asking open-ended questions. Open-ended questions start with "where" ("Where did you come from?"), "how" ("How do you like it here?"), "what" ("What do you do?"), or "which" ("Which department are you at?"). Open-ended questions get the conversation started.

To be effective at exchanging information, you will need to build a relationship

---

It's better not to ask "why" ("Why are you here?"), since that could be perceived as an attack, especially if you don't know someone. You want the other person to be open to communicating. Also avoid closed-ended questions that can be answered with a simple yes or no. So don't ask things like: Did you think it was boring? Isn't the coffee awful? Instead ask "What did you think about that meeting?" or "What do you think about the coffee?" You can also express your own opinion, for example, "I think the coffee has been a bit weak lately. What do you think?" If the other person responds, you've made the initial contact.

At this point, you may want to continue the conversation by asking a question focused on content, such as "What are you working on?" or "What did you think of the speech?" That's all there's to it. Just go through the steps.

Making small talk may feel awkward, but once you master this skill, you can get to know more people and build relationships. Go ahead and practice—you will see it's easier than you think. And most people will appreciate that you use small talk to make contact with them.

Leadership

Stakeholder management, what does it mean?

An engineer asks:

I often have a hard time getting my ideas and proposals accepted. My manager says that I need to involve my "stakeholders" more during the decision making process. Who are they, and how should I handle this?

The communication trainer answers:

Stakeholders are parties that have a stake, or interest, in your solution. Stakeholders may be part of your organization, or they may be external, for example suppliers or customer. Ask yourself: Who has an interest in your work? Who will be impacted by your solution? Who influences your work? This is how you find out who your stakeholders are.

Next, create a matrix for your stakeholders. The y-axis represents the amount of influence they have, and the x-axis shows how big their stake, or interest, is. Train passengers are a good example of stakeholders with little influence and a lot at stake. For them, it's important that the train is clean, there's a seat available, and the train leaves on time. But do they have any influence when they're waiting and the train isn't coming? None whatsoever. In some countries, train passengers have organized to advocate for their rights as travelers. Unions arise for similar reasons.

When it comes to your solution, there may be a lot at stake for the production and customer service departments, for example. But they might have little influence and will be facing the consequences at the very end of the process. So make sure to involve these stakeholders at an earlier stage, or at least keep them up to date.

Management may be a stakeholder with a lot of influence and no immediate stake or interest in your solution. Here the motto is: "Keep them happy." It's important to understand what matters to them and to get their support early on. Make sure your proposal is in line with their plans. Let's look at an example of a small team that needed to solve an urgent problem. Team members went right to work and sent a few emails to inform others who were involved in the project. By the time they presented their solution to a large room full of stakeholders, they had already ordered the hardware. But guess what. A manager in the audience calculated the project's cost. He decided to stop the project on the spot and left the room. Forgetting a stakeholder may lead to unwelcome surprises.

> Forgetting a stakeholder may lead to unwelcome surprises.

Some stakeholders have a lot at stake and have a lot of influence. These are the people you need to work with closely to get the project done. You probably won't forget them. They are the leaders and managers of the parties that are directly involved, suppliers, and teams you interact with directly.

Finally, there are stakeholders who have little at stake and little influence. You can ignore them. Or better yet, monitor them, because some stakeholders can become important as the project progresses or because of organizational changes. Just be clear which stakeholders you need to target, because after all, you don't have all the time in the world.

Once you know who your stakeholders are and what their influence and interests are, find out to what extent they're aligned with you. The stakeholder who is most affected by the problem you're solving, will be most supportive. They can support you in persuading others who have different interests. The requirements for the optimal solution will flow directly from the interests the stakeholders represent. For example profit margins, time-to-market, product quality, a solution's reusability, technical risks, a technique's proven track record, etc. Some criteria may be less technical but equally relevant, such as the company's branding, user-friendliness, or design. Present a limited number of technical solutions, three at most, and work with all the stakeholders to decide which one is in the company's and the customer's best interest.

How do I get people to really listen when I present a solution?

A software engineer asks:

I regularly meet with a team of stakeholders who need to make decisions about the solutions and recommendations I present. The problem is that a few minutes into the presentation, I usually get sidetracked by their critical questions. How can I get them to engage from the very beginning and make sure they reflect and really listen?

The communications consultant answers:

First and foremost, you will need a substantive argument if you want to convince and engage a group of people. It doesn't matter how flashy your presentation is—engineers want facts, not fiction. So come well-prepared and know what you want to say.

Let's assume your presentation is solid in terms of content. It can still be difficult to convince others. Listeners may disengage right at the beginning, or you may not get a chance to tell your story because they immediately start asking lots of critical questions.

In those first few minutes, your main goal is getting your audience to be willing to really listen to you. You achieve this not just with the content of your presentation, but by delving into the interests of the people at the table. Oftentimes, they are stakeholders, coworkers who have their own interests in the project. What do they worry about? What are their plans? What is important to them? In the first few minutes of your presentation, everyone wants to understand why the topic is relevant for them and get the sense you understand them.

How do you do this? An engaging introduction for your presentation consists of three steps.

Step 1) Clearly describe the problem and the consequences of inaction. What is the crux of the problem? The root cause of the problem might be a highly technical issue that not everyone in the meeting will fully understand. If that's the case, you will have to start by translating what the technical problem means in terms of its consequences for the stakeholders at the meeting. For example, a technical failure could result in late delivery. That is bad news for the project manager. Or a delay will cause problems in the field, which will need to be solved by Customer Support. Describe exactly what the problem is and quantify as much as possible, using terms that will engage your listeners. This will get their attention and provide the necessary motivation to take action. Know that if it doesn't hurt, nothing will happen.

Step 2) What do you want to achieve? For example, your main goal may be to deliver on time. Or maybe you want to reduce the number of problems in the field. This is the common goal everyone can agree on.

Step 3) Indicate what you want to achieve with the meeting. For example, your goal may be to brainstorm possible solutions or to decide on a specific approach you prepared.

Engineers want facts, not fiction.

These three steps will ensure that all the stakeholders at the meeting will be interested in what you have to say. After all, they recognize the problems and objectives you describe. This is important, because we're not always willing to acknowledge a problem. Say you are on your way to the dentist. How is your toothache doing? Chances are, your tooth isn't hurting quite as much. So what will your dentist do? Your dentist will probe your diseased tooth to let you feel you do indeed have a problem. The result? You are once again willing to work towards a solution, and your dentist will tell you what your options are. This approach is similar. After an engaging introduction, your presentation will be more convincing. And your listeners will be more open to really listening to your story.

As a specialist, how can I take on more of a leadership role in the organization?

A lead engineer asks:

Over the past few years, I've worked my way up to the position of lead developer. My technical knowhow is solid and widely recognized by the people within the organization. Now they expect me to take on more of a leadership role. What should I focus on?

The communication trainer answers:

Performance, People, and Process are key. Start your development by mastering the first of these three P's: Performance. You have to learn how to do your work well. This will give you the self-confidence and credibility to make your way into the organization. The other two P's will come into sharper focus next: You should be able to work with people outside of your own projects, know how to engage stakeholders, and convince others with your ideas.

This step starts by acknowledging that you are not operating on your own. You are part of a team of people working together to achieve results. In other words, you need to learn how to take in a 360-degree view. Look at who's "above" you: the various line managers, project managers, business managers, and other stakeholders who influence the project or have a stake in it. Look at colleagues at your own level who have similar roles in adjacent technical fields (e.g. other lead engineers). Also look at coworkers in positions "below" yours: the people you manage with regard to technical matters, delegate work and responsibilities to, and help by sharing knowledge and experience.

In order to engage and convince people, it's important to know what motivates them. What is important to them? This requires that you learn to see things from their perspective. Once you know what's important to the other person, find out how your ideas can help them. How will the technical solution you have in mind solve their problem?

The project manager who is "above" you would probably like to make the deadline. This is a short-term interest. His long-term interest could be for the project to be done well to avoid needless errors from arising after project completion. This would be costly and stressful for the organization and damage his credibility.

As for colleagues at your level in an adjacent field, you will need to try learning about the technical foundations of their area of expertise. If your colleague is an optical specialist, for example, it means you will have to take a crash course in optics. Once you see how your work impacts the work of others, it becomes clear what you need to discuss. Only then you can develop a critical view on the existing process and start a dialog.

Performance, People, and Process are key

For your own development, you will need to delegate work and responsibilities to coworkers in positions "below" yours. It takes time and mental bandwidth to grow your sphere of influence within the organization—it's not something you can just do on the side. If you are overwhelmed by the details of your work, you may not even notice the stakeholder standing next to you at the coffee machine, let alone know what message to convey or what to ask.

Try to delegate work in such a way that the other person can learn something. Focus on sharing knowledge and allocate time for providing guidance. The overall effect will be that employees will enjoy working for you. That way, both you and the people around you will grow.

How do I make decisions based on incomplete information?

An engineer asks:

Recently, I switched my engineering job for the role of architect. Now people more often ask me for my opinion on the direction of a project or on how to tackle problems. When I know something for sure, making a decision or giving my opinion is easy, but it's harder when I'm not 100% certain. In that case, I tend to procrastinate or prefer to look up some more information. But I usually don't get the time to do so. How should I handle this?

The communication trainer answers:

When you go from strictly being an expert to taking on a more technical leadership role, for example as an architect or a principal engineer, you will increasingly be asked to take decisions with incomplete data. Designers and many more stakeholders, such as managers, project leaders, salespeople, and suppliers, will all want to know your opinion. You will not be able to avoid this—it is part of your job.

Normally, you may only want to make a firm statement when you are a hundred percent certain of something. But you can't hold up the project. So what happens, for example, if people ask you for your opinion during a meeting, you know something with eighty percent certainty, and you hesitate or don't say anything? Someone who is more assertive may have only thirty percent of the knowledge but will nevertheless put his or her stamp on the project and decide its direction.

This is where you need to assume responsibility as a technical professional and take the lead. Don't do this by bluffing. That doesn't feel right and most people will see right through it. You need to take a stand. Say what you think should be done. Do this by using phrases with "I," for example: "I think we should do this and that." Using the word "I" adds authority, clearly states your position, and makes you visible to others. Now people can follow you (or not, of course).

Simultaneously start researching any uncertainties

You can also point out any risks or unknowns you see and describe an approach to manage these. A common approach is to take a decision and simultaneously start researching any uncertainties. If it turns out your solution was right, you haven't lost any time. Should the new information reveal you didn't make the right choice, you can still make a course correction.

You want to be one hundred percent certain partly because of the fear of making mistakes, and partly because that's how you were trained. Learning to make choices without knowing everything, starts with the realization that it's necessary. Next, resolve to take more frequent, timely decisions. It's important to go beyond analyzing the available facts: You have to learn to start trusting your own experiential expertise and that of others.

Many people call this as a "gut feeling." However, learn to distinguish between the gut feeling caused by last night's hamburger, your own uncertainty, or the feeling that says something is right or really wrong. For example, if something feels wrong but you can't quite put your finger on it, it's worth mentioning this. For example, you could say: "I'm not sure what it is, but this solution doesn't feel right. What do you think?" Often someone else will know exactly what the issue is. You don't have to do it all alone. Ask the experts you trust for help. You can figure it out together.

In short, as a technical leader you need to learn how to make decisions, ask others for help, and rely not just on facts but also on your own experiential expertise and that of others. Go a step beyond being a scientist and truly take the lead.

What does the board focus on when I present a solution?

An architect asks:

I have to present my proposals to the board of directors on a regular basis. During these meetings, I show them what I think is the best way forward. After a presentation I often wonder what I could have done better. What does the board focus on most?

The communication trainer answers:

The most important thing for you, as a presenter, is to understand that the board has to make decisions based on your presentation, about whether or not they should make an investment for example. In other words, the board members will need to make up their minds based on the information that is communicated in a half-hour window. In essence, they need to be able to trust that you and your team are doing the right thing. The board members will need to base their decision on the limited information you can provide in the limited available time, and on the way you and your team do that.

Obviously your proposal needs to make sense in terms of content, but there are a number of other important factors:

1) Is your team presenting one story, as one team? If not, this could point to disagreement within the team, which in turn could indicate a problem with the content.

2) Are you "congruent" as a presenter? Meaning, does your body language match your words? The well-known statement that nonverbal signals have an enormous impact on communication, is especially true when the content of your story is not supported by your body posture and your voice. Say you make a firm statement during a presentation while nervously plucking your pants. Your message will come across as less certain than your words alone suggest. This would certainly be a reason for the board to ask probing questions. The same thing happens when a team member starts looking uncomfortable during a particular part of the presentation. These are all reasons for the board to start grilling you and to find out if there's more to the story.

3) In terms of content, it's important to use actual numbers for your presentation. Avoid tables with pluses and minuses, smileys, or other qualitative indicators, and present real numbers instead. This prevents personal discussions (my "smiley" is better than yours). Using real numbers will support your presentation and allows board members to truly compare different options. Create a reliable decision table with your team members, so you can present the best possible input. It's best to compare three options. Why three? Presenting one option doesn't really provide a choice, two options call for a third option, and when there are four, the first one has already been forgotten. Three options will provide just the right amount of information to make a choice without causing confusion. Compare options using relevant business drivers, such as costs, performance, risk, planning, etc.

4) Always indicate which solution you prefer. Take a clear position, explain your thinking, and put your full weight behind it. Also be prepared to let go of your position if new facts or more important considerations are brought up. What matters is choosing the solution that works best for the company and the customer.

5) Prepare the right back-up slides in case you need more details. When you have them, you usually won't need them. But when you need them and don't have them, you have a problem.

6) Should you act more certain about a particular solution than is warranted? No. Honesty is the best policy. People will find out sooner or later if you are bluffing. Sometimes people can tell right away, because you're not congruent (see item 2). Or they may find out later that your statements were not fact-based. Do take a stand, and do it with conviction. If there are uncertainties in your story (and when is this ever not the case?), it's fine to mention them. Also mention how you are going to handle these uncertainties. Use words like "I would propose" or "Our team thinks the best solution is to..." Take the lead, so others can follow.

Simply say you don't have the answer right now.

---

7) Don't know the answer to a difficult question? There's no point in telling yourself you should be able to answer every question. That adds unnecessary pressure and could cause you to shut down. Simply say you don't have the answer right now ("I don't know") and you will get back to them within a couple of days. This shouldn't be a problem, unless you should have known the

answer as part of your job. And maybe a colleague can provide an immediate answer, providing a nice opportunity for collaboration.

As a final tip, always try to prepare an important presentation by practicing in front of your coworkers. This will give you a chance to fix weaknesses in your presentation before the real work starts.

What should I do when someone attacks me?

An engineer asks:

Our team is under a lot of pressure to make the deadlines. Management also has an interest in the project, which means we have to report on our progress every week. Because of all this pressure, we often end up dealing with both coworkers and management reacting pretty aggressively. I would like to handle this the right way. How should I do this?

The communication trainer answers:

Being confronted with aggression is no fun. It's like a slap in the face, and you're startled, or maybe you want to go on the offensive. When we think of aggressive people, we might think of people yelling or throwing beer glasses. One dictionary defines aggression as "someone attacking another person," and that includes verbal attacks too. These types of attacks occur more often than you might expect, especially when people are under a lot of pressure.

When we study aggression more closely, we see two types of aggression that need to be handled in different ways: frustration-aggression and instrumental aggression. Frustration-aggression is characterized by the emotions that are involved for the person attacking. A project leader can become so frustrated about project delays that he becomes angry and says things he may regret later on. Emotions take over, and he is no longer able to listen effectively. At that moment, only one version of reality exists in his mind, and most of his perception is colored by anger.

Daniel Goleman, who came up with the term "emotional intelligence," says that when you are in an emotional state, 90 percent of your perception is determined by the emotion. In other words, you only see 10 percent of the facts. This means people will need to calm down before they are approachable and can revisit the situation with an open mindset.

That's why frustration-aggression is handled through de-escalation. Someone will need to let off steam before he or she is able to listen again. This is what you do:

1) Stay calm. For example, take a few good deep breaths, and take a step back from the situation. Don't argue and be flexible instead.

2) Invite the other person to tell you what exactly is bothering them. For example, say: "Tell me, what's going on?"

3) Listen and show that you understand what they're saying. You could say something like, "I get why you would be upset."

4) Ask follow-up questions until the issue is clear.

Keep responding mindfully instead of getting angry.

Instrumental aggression on the other hand, means that someone is consciously using aggression to reach a particular goal. They may raise their voice or pressure you. How do you handle this? Focus on establishing boundaries for the other person. You need to be very clear that this behavior is not acceptable to you.

This is how you do it:

1) Remain calm.

2) Tell the other person that you are really bothered by the way they're talking to you, and tell them what you don't want them to do.

3) Show them that your intentions are good, as long as you're treated respectfully. For example, say: "I would like to solve that issue too, and that's why I suggest we calmly discuss the matter."

4) Work together to find common interests and a way to resolve the situation.

Note that someone with functional anger may have a legitimate point. Maybe you're not putting in enough effort, and you needed that slap on the wrist. So when you're dealing with aggression, it's important to keep responding mindfully instead of getting startled or angry. An important practical tip is to always pass back the ball. Ask the other person, "What do you mean exactly?" You will get more information about the reason behind the attack, and you have a moment to regroup and get an overview of the situation in a relaxed way.

How can I quickly align with my key stakeholders?

An architect asks:

As a systems architect, I have to work with various stakeholders, but I tend to forget about them, especially senior management I don't run into every day. I don't always involve them on time or give them due consideration. That is detrimental to me and the project. I have been called on it a number of times, which is frustrating. How do I properly coordinate with senior stakeholders without investing a lot of time?

The communication trainer answers:

First of all, it's important to realize stakeholders aren't just stakeholders. They are people made of flesh and blood. They will consider certain things important, have their own worries, want to be informed, and want to feel in control. To coordinate properly, it's important to put yourself in their shoes and find out what's important to the stakeholders you work with. This is strongly content focused. But you also have to be able to connect with someone and size them up. How do you do that?

A simple model provides some insight into a person's character by using two axes: an introversion/extroversion axis and a factual/relationship-feeling oriented axis, resulting in four quadrants. For introvert-factual ("Conscientious" in this model), you can typically think of accountants, mathematicians, and technical people. They are strongly detail oriented and good at following procedures and methods. Someone who is more extroverted and factual will be more externally oriented. He or she will be focused on actual results: "When will this be done?", "How much does it cost?" etc. Think of a manager, for example ("Dominance").

Extroverted and more relationship-oriented, brings us to the "Influencer." "Influencers" are focused on new ideas, working together, and visions for the future. As you've probably guessed, this creates tension with a conscientious person who is more risk averse. In turn, the Influencer will quickly experience any talk of risk as an obstacle. A more introverted but relationship-focused person is called "Sociable." This person will help others without boasting about it and is the glue that holds the team together. Someone who is Sociable can easily be overruled by a more businesslike person, often the manager ("Dominance").

When you look at personality, most people have two or three well-developed aspects and one aspect that is clearly less developed. Now, let's get back to those senior stakeholders. They are likely to have some "Dominance" as part of their profile. Otherwise, it would be hard for them to be successful at their job.

> *It is always appreciated when you put energy into coordinating*

What is the best way for you to approach them when you try to get their support? The easiest way is to be clear and concise. Keep it short and to the point, for example: "I would like you to sign off on something" or "I'm dealing with this problem and would like to handle it this way. What do you think?" It's important for you to tell them in a few sentences what you want their input on. Managers like this, because you give them the opportunity to manage without any fuss. Generally, you will get a response right away. This could be anything from "Fine, go ahead," to "We have to talk this over, let's set up a meeting," or "Discuss it with so-and-so first." In any case, you will be one step further. Coordinating can often be done in a minute. However, it is crucial to be prepared. This is not something you want to do on the fly.

What is the best way to get a hold of the stakeholder? You could try to make an appointment, but that may be difficult. Be smart about it. For example, go to a meeting you know he (or she) will attend. Approach him during a coffee break. You can also find out when he is at his computer responding to mail (for example from 7:30 to 8:30 AM). Make sure you are waiting at his desk at that time. It will be appreciated when you take the initiative and put energy in coordinating, because it makes managing easier. He's happy, and so are you.

How do I deal with the vague suspicion that I'm being deceived?

A systems architect asks:

A large part of my work consists of reviewing the work of others. During those reviews, I often get the sense that people are making the numbers look better than they are, manipulate results, and keep information from me. What's the best way to respond and still reach the right decision?

The communication trainer answers:

Whether information has been distorted, omitted, or manipulated will often not be clear until later. At that point, you will probably realize you already had some suspicions. For example, someone may present a proposal as if everything is clear-cut, and you wonder if that's true. Or someone presents a problem in a way that indicates there is only one possible solution. But was the problem analyzed correctly? You may have some vague notion that something isn't quite right or something important is missing.

Here are two wrong ways to respond. The first is to ignore your intuition and let things go. This often happens when your mind is busy. You can't hear that little voice in your head that tells you something is wrong. Maybe you're thinking about the next meeting or worrying about something else. Even if it does reach your consciousness, it's easily pushed aside because so many other issues demand your attention. In hindsight, you may realize that you sort of knew but didn't act on it.

The second wrong response is to get focused on interpretations or fantasies. You notice the presenter uses certain assumptions as if they are true, and you start to think there must be more behind it. For example, you may think: "Her goal is probably to do this or that, and she ignores A and B so we'll all think that can't happen, etc." This brings up all kinds of emotions that make the story even bigger. You end up with a series of interpretations. It may very well be true, but you don't know for sure. What you do know, is that it stresses you out. Many studies have shown that your perception of reality quickly diminishes when it's influenced by emotion. And if you respond from that emotionally charged interpretation, chances are that it will evoke an emotional response in the other person as well. You've made a mountain out of what might be a molehill, and no one even knows what it was all about.

How can you do justice to your suspicions that something is wrong, and at the same time quickly clarify what's going on? You do this in three steps:

1) Always try to keep a calm mind, and make sure you're not distracted. Are you reviewing someone's work? Focus on the review. That way, you are in the best position to notice if something feels "off."

2) Briefly investigate your gut feeling. Does it have to do with the topic at hand, or with the heavy meal you had last night?

Address your initial observation, not your interpretation.

---

3) Does your gut feeling relate to the topic at hand? Mention that right away. Has your mind already started interpreting? Here, you have to be disciplined by communicating your actual observation rather than your interpretations. For example, you can say: "I'm noticing that I have doubts about this assumption" or "I'm wondering what it is exactly that you're asking us." So talk about your initial observation, not your interpretation of the possible reasons behind it. Also make sure your voice doesn't reveal any irritation or distrust. Just express your concerns and be curious.

Other attendees will now be motivated to take a moment to reflect on what you're saying. This is how you create the best conditions for an honest conversation about what is going on. If someone still tries to deceive you by telling nonsense, you can still put more pressure on them and go on the offense. But usually, that won't be necessary anymore, and that saves you a lot of energy.

How do I get the organization back in line?

A team manager asks:

I'm losing grip on my team. They are missing deadlines, do many things without my knowledge, and are coordinating much of the work with project managers without involving me. Team members aren't collaborating well and often fail to honor their mutual agreements. In other words, I feel like I have been sidelined. On top of everything, I'm held accountable for regular project delays and low quality work. I'm at the end of my rope.

The communication trainer answers:

If you're no longer in charge of your team as a team manager, you might wonder if it's time for a career change. But it would be more interesting to ask yourself what caused the problem and how you can solve it. Your responsibilities as a team manager are roughly threefold: (1) ensuring that your people can be successful at the projects they're working on; (2) ensuring that your team has the necessary capabilities and skill level, and (3) ensuring that your employees have development opportunities. In the meantime, you also take into account the various stakeholders involved.

In your situation, it sounds like you are turning the knobs, but nothing is working. You have lost control. Project managers are putting people to work on projects without consulting or informing you, thereby robbing you of your most valuable assets: your employees. You will be unable to adequately fulfill any of a team manager's responsibilities. In order to become effective again, you will need to reassert your position.

Get your project managers to agree that they can only assign work to one of your employees based on what you decide. Project managers will tell you what they want or need and who they are thinking of, and you will indicate who is available and when. Together you will look for a suitable solution. Of course employees will have a big say in this too. The three of you will put your heads together, so to speak.

First you will need to tell project managers very clearly that they can no longer bypass you—they will need to work with you. If they don't, they will have a problem. Second, you will have to explain the new work process to your employees and clarify that you will be holding them accountable. For this entire process, it can be helpful to gain senior management's prior approval. Do this all in a well-organized manner, and you will regain control so you can go back to deciding who is doing what, when, and for how long. Now you can tackle other tasks too, such as developing competencies.

Clearly, this is a matter that can only be resolved by and among you and your colleagues at management level. It is a "Chefsache," a leadership issue. A non-committal conversation with a vague outcome will be counterproductive in this case. What matters is who is in charge and who is not.

It's "Chefsache", It's a leadership issue

That's why you will need to determine ahead of time how much changing the situation is worth to you. Can you afford to let this go? If every fiber in your being tells you things are not going well, you will have to commit fully and let the chips fall where they may. That's the only way you can be authentic and give it all you've got. Whether you will succeed is hard to predict: How much does your company need you, what office politics are at play, and how much credibility do you have?

A 1980 soccer game between two Dutch teams nicely illustrates what a "leadership issue" is. Soccer legend Johan Cruyff was watching from the stand and could no longer bear to watch how poorly his favorite team, Ajax, was playing. Cruyff got up and took the lead without asking the official coach for permission. His forceful intervention turned the game around. Ajax went from losing 1 to 3 to winning 5 to 3. Needless to say, this is not something you want to do every week. Choose your battles.

How do I deal with ideas that are treated as "sacred"?

An engineer asks:

As a systems architect, I often deal with experts or stakeholders who are deeply invested in a particular technical solution. In other words, they put so much faith in a certain solution, that they're not open to the facts I present. How do I handle this and let the facts speak for themselves?

The communication trainer answers:

The software in our brains is programmed for us to have a bias or preference for anything that confirms what we're already thinking. It's stressful when we see or think something that is not aligned with our beliefs. This is called cognitive dissonance. It turns out that when we get good news while expecting bad news, our initial response may resemble disappointment, for example when a doctor tells us we're fine while we were expecting bad news. We are strongly attached to our world view and don't like to see it disrupted. As a result, we don't see the facts, we don't want to face the facts, or we simply don't accept the facts. Looking at the same facts, we reach the opposite conclusion. Is the glass half full or half empty?

How can you let the facts speak for themselves? You will need to present as much factual evidence as possible, regardless of any taboos or fear of the possible outcome (maybe that great idea wasn't so great after all?). This may involve a lot of homework. Next, don't just present your conclusions—present all the raw data your conclusions are based on too. Be as clinical as you can be. Think of a doctor in wartime. He has one serum, and he needs to give it to the wounded soldier who has the best chance of survival. The doctor tries to think in a purely analytical way, without any personal preferences. Such a "cold-blooded" approach will allow you to come up with the best assessment of opportunity and risk. And don't be afraid to take a critical look at your own ideas. Then, act based on your understanding of reality. It is important that this understanding is as accurate and complete as possible; otherwise, the outcome of your actions will be suboptimal at best.

> **Make an effort to find sources that you wouldn't normally consult.**

The fact that your actions are based on your conviction also makes it more interesting for others to try to influence your understanding of reality. People can make a conscious, often calculated decision to present or "frame" information in a way that leads to a particular interpretation. For example, "the glass if half full." This is true, and it can give you the sense you don't have to go out for milk yet. "The glass is half empty" is true too, but it might motivate you to run to the store right away. The fact is, the glass is 50% full and 50% empty. This phenomenon of presenting, hiding, twisting, etc. information a certain way, happens all around us. How can you protect yourself from this?

Look at an issue from different angles by always consulting different sources. To prevent blind spots, make an effort to find sources that you wouldn't normally consult and that may even invoke some feelings of aversion. Try to be broadly informed. Look for real facts based on experience. Are you a manager? Chat with the people who work with the machines, and ask them for their opinion. What do they think about the new solution? What type of problems do they expect? It's important to get all the facts out on the table, no matter what.

What should I do when no one wants to listen?

A specialist asks:

I really don't agree with the direction a project is moving in. I have spoken up, but nothing changes. This concerns me. The situation doesn't make me any friendlier, and people around me seem to agree. Nevertheless, I'm convinced that I am right. How should I handle this?

The communication trainer answers:

Being right and being convincing are two different things. The latter can be difficult when your message is not in line with the dominant opinion or people don't see the risk or problem you point out. You can persist and keep making your point, but this will impact group dynamics. Soon, you will no longer be taken seriously: "Here he goes again..." Socially, it isn't smart to keep repeating yourself. You will become an outsider. How do you prevent this from happening?

It takes courage to state an unpopular opinion. The trick is to do it in such a way that your point isn't lost and you remain standing. There are different ways to do this. Look for people in the organization who will support your vision. It can make sense to escalate the problem you've identified to higher levels in the organization. They should have a broader view of business operations and should be able to correctly value a different opinion.

Is senior management not responding either? Maybe your timing is off. You may be too early. The motivation to listen often emerges when a problem is experienced in real life. You can assume that seniors within the organization are alert to future risks. But even at that level, day-to-day problems often determine what's on the agenda.

Don't stress about things you cannot change

The most serious risk is that you get fed up, lose your motivation, and start to complain, thereby digging your own grave. Don't do it. Try to stay optimistic and trust that things will work out.

So what can you do? One tactic is to wait for the moment the first symptoms of the problem you've been trying to address become evident. Let's say it concerns a part you don't trust. Are there any measurements that prove your point? Have any of the parts broken on the test bench yet? Have there been any complaints in the field yet? If this is not the case, the only thing to do is wait.

Maybe you can put a little effort into working towards a solution, in case the potential problem does occur. The trick is to focus on what you can do and not stress about things you cannot change. This way, you remain proactive and solution oriented.

Team management

What's the best way to delegate work?

A test engineer asks:

I'm often saddled with projects I hardly have time for. My manager says I should just do a better job delegating. But how do I do that? I don't want anything to go wrong.

The communication trainer answers:

It's important to get it right when you delegate work. If you do it wrong, you end up with a lot of extra work, often at the very last minute. The person you delegated the work to would also experience it as a failure. You may even feel justified in thinking that you're better off doing things yourself to get them right, and that would be the worst outcome. Of course it would be faster to do it yourself. But if you keep doing the work yourself, it will hamper your own development as well as your coworker's development. So delegate, but make sure you get it right. How do you do that?

If you want to delegate, you will first need to decide what you want to delegate. Choose a bite-size chunk of work, a clearly defined task that allows someone to fill in the details. Go ahead and tell the other person what you want, and then let him figure out how to do it as much as possible. That way, you give the other person greater responsibility for the implementation. He can do the work and take credit for it too.

The first conversation is the most important moment of the delegation process. The outcome of this conversation should be a clear agreement about what needs to be done and how you will structure the process. Think about the amount of guidance you will provide and how you'll handle any questions that come up. For example, can you be reached by phone anytime, or would it be better to compile a list of questions? Don't forget that some people prefer to ask questions before thinking things through, while others are reluctant to ask questions and may keep going around in circles for too long. Make sure to discuss this.

The agreement should also clearly state what the deliverables are, when they are due, and what format should be used (e.g. a table or big stack of paperwork). Also check the level of commitment. Does the other person really want to do the work, or do they feel they have no choice in the matter? The best way to find out is by asking an open-ended question: "So, what do you think?" Pause and pay attention to body language. If you notice someone is looking away or speaking with hesitation, don't let it slide. Delve a little deeper and say, for example: "I see you're hesitating. Do you have any other questions, or are you uncertain about something?" This way, you get to the bottom of things. Better to do this now, while it's still manageable.

> **The first conversation is the most important moment when delegating**

Let the amount of guidance you offer depend on the person's experience with the particular task. We call this "task maturity." It's quite possible a task is new for someone, even if they have 20 years of work experience. In that case, you would need to sit down with them more often than you would with someone who has a lot of experience with a particular task.

Others may have some experience but haven't completely mastered the task (even if they don't realize it). Give them a chance to make some mistakes first, so they can continue learning. When the person you're delegating to has a lot of experience with a particular task, it's best to let him decide—as much as possible—how to approach the work. Too much guidance would be experienced as patronizing and demotivating.

Try to meet on a regular basis to review progress while the project is ongoing. This way, you can make timely course corrections, and the confidence in a successful outcome will increase. And that's a good thing, for you, the other person involved, and the company.

How do I become a good mentor?

An engineer asks:

My manager says I have to transfer my knowledge to open up new career opportunities for myself as well as for the members of my team. Simply delegating work is not sufficient. I have to train them. How should I handle this?

The communication trainer answers:

If you want the people in your team to develop, you will need to mentor them. That means mentoring will become part of your job description as a technical leader. You'll have to pay attention to a number of things to be an effective mentor. First, it's important to realize that in your role of mentor, your primary focus is not to solve the technical problems your mentee is working on. Instead, focus on your mentee's development. You do this by teaching them how to work towards a solution of the problem. In practice, this means that you support the mentee, whatever happens.

Create a "safe learning environment" before you start sharing your feedback. This doesn't need to take long. Always start by briefly checking in and connecting with the other person, for example by asking: "How are you?" Then make the switch to discussing the work. You could say something like: "Let's take a look at your work. Is that OK?" The mentee will respond "OK." Now you can give your honest feedback. Your aim should be to continuously help the mentee master the content. Being too focused on the problem itself won't provide a solid foundation for a successful mentorship.

Another success factor is the quality of your feedback. Feedback is key to development. Getting good at giving high-quality feedback is something you'll need to work on. What does this mean? Make sure your advice is concrete. Now, the term "concrete" may sound vague, so how can you tell if you're "concrete"? Here's a tip: Ask yourself if you can visualize it, if you can see it like on a TV screen. Compare "Make sure your story is short and specific" to "Tell your story in six slides or less and use numbers whenever possible." Don't immediately assume your mentee will understand what you mean, because that might not be the case.

To get your mentee to take your advice and actually put it into practice, you will need to provide insight in what does or doesn't work in their approach. Make this the focus of your feedback. Address the mentee's actual behavior and the effect it has, but be as specific or "concrete" as possible. Is the effect positive? Mention that and compliment them. This will motivate the mentee to continue doing the right thing.

Be as concrete as possible to help your mentee.

Also describe the negative consequences of their actions to show why a certain approach doesn't work. This causes a bit of "pain" and will motivate the mentee to find out how to do better. And you'll be right there to discuss a way forward or to give advice. Let's look at an example. You can say: "I saw that you presented thirty slides with many details that most attendees could not understand. The effect was that many people lost interest." The mentee responds: "Ouch... OK, I see. How can I improve?" At this point, the mentee will want to pull information from you to improve, while more senior mentees will often be able to come up with a solution themselves. This way you won't have to push your advice.

If you're trying to be a super-friendly mentor and don't say what went wrong, you're not helping your mentee—you deprive them of opportunities for improvement. And you will end up pushing your solution.

So be as concrete as possible to help your mentee, and balance positive and negative feedback. Start transferring your knowledge this way, and it will have a positive side effect: Someone who can learn from you will be motivated to work for you and to continue working for you in the future. That's how you can create a team of people who will be happy to work for you. You don't have to pull rank on anyone, and you create space for your own continued development.

How can we become a strong project team?

A project leader asks:

As a small project team, we are responsible for managing an important project. Management is steering us in a direction that doesn't feel right to us, and this has happened a few times before. Team members often disagree with each other too, and the result is that we can't get it together and start lagging behind. How do we turn the tide and get a handle on the project?

The communication trainer answers:

The problem is that you are not working in unison as a team. And since a forceful, unambiguous answer is lacking, your team can easily be passed over and sidelined. Stakeholders can overrule you with little effort. This usually doesn't benefit the project. After all, your team knows best when it comes to the project and how it needs to be handled.

The first step towards a solution lies in your collaboration as a team. Of course it's always possible that something changes, like the circumstances surrounding a project, the company strategy, or a customer's needs, or maybe colleagues working on a related project develop new technical insights. All of this can affect the course you decide to take. But if you don't take a stand together as a team, you will lose control of the project and might make the wrong decisions, which makes managing the project very messy. How do you gain control over this process?

Two factors are crucial to your team's ability to manage the project effectively. First, make sure that the team has the capabilities to manage the project's most important aspects, such as cost, quality, time, and staff. This allows for the right kind of discussion, with team members working together to examine all these aspects and propose the best solution for the customer and the company. Every team member should be able to fulfill their role in that discussion.

This leads us to another important aspect to improve the strength of the team. A supportive environment is important for meetings to run smoothly. This means that people trust each other and aren't afraid to speak their mind. They listen to each other and are allowed to make mistakes or voice different opinions. If you manage to achieve that, you will have effective meetings, and each team member will feel involved once decisions are made. This puts everyone on the same page when it comes to doing the work.

> **The first step towards a solution lies in your collaboration as a team.**

When the project is under pressure to move in a different direction, it's important to figure out where the team stands. Can we live with this? Yes, if certain conditions are met, or no, unless... Or maybe the answer is simply a wholehearted OK or Not agreed. You need to come up with a clear proposal about what you all think is the best way forward. Next, it's important to be unified when you communicate your proposal to various stakeholders. This is how you influence the stakeholders you work with. Make sure you all tell the same story, so the whole team conveys a clear, well-reasoned opinion. Work together with the stakeholders to set the final course.

How do I get a quiet group to talk?

An engineer asks:

I often conduct meetings with a lot of junior employees, but I feel the group is not really responding enough. They don't say much and take a wait-and-see approach. It's mostly just me talking. How do I encourage active participation and productive discussions without losing control of the process?

The communication trainer answers:

Picture a meeting with the chairperson talking and everyone else slumped in their chairs—disengaged or thinking of their to-do list. You know this is a waste of time. But how do you start a meaningful conversation during a meeting when no one says anything? In this case, it is a disadvantage if the chairperson is good at "taking the floor." So should he or she be quiet? Sometimes, yes. The art is in knowing when to speak and when to remain silent. So how does this work?

The person leading the meeting will need a number of soft skills to get a meaningful conversation started. Let's say the chairperson has just raised a topic during a meeting and wants people to have a good discussion about it. The first pitfall presents itself right away. When she leans in expectantly and looks at the others, they are still leaning back in their chairs in "listening mode." Nothing happens. What just went wrong?

By leaning in and actively looking around, the leader's body language shows she is still active. Others will often subconsciously respond to this by being passive. They don't "have to" do anything, even if the chairperson politely asks them: "What do you think about that?" To elicit an active response, she will need use her body language to show that she now expects others to talk. She does this by becoming passive. So when the leader asks "What do you think about that?" she leans back, relaxes, and says nothing for at least four seconds. Four seconds provides enough time for everyone to process the question and feel that dreaded "prolonged silence" arising.

The chairperson's challenge is to stay quiet, relax, and look around without staring. Now someone in the group will respond. The person responding will direct his or her response to the chairperson.

As a chairperson, you will have to ask follow-up questions until it's clear what the other person means. Summarize the response ("Okay, so you're saying that..."). Next, pass the ball ("How do others experience that?"). Look around in the meeting in a relaxed way. Someone else will respond to that. At this point, be aware of another pitfall.

Taking back control may be easier than you'd expect.

If you keep responding as the chairman, the discussion will take place through you, and participants won't start discussing the topic with each other. So when someone responds, use your body language to "pass" their response on to another person in the meeting. Use your eyes and shoulders: Take a moment to show that you've heard the response, then look away in the direction of someone else. The person who responded will follow your glance without even realizing it and focus on this other person in the group. That person in turn will feel called to respond. And all of a sudden, the group is talking. Now all the chairperson has to do is relax, listen, and keep the conversation on topic.

If necessary, you can play devil's advocate to create a dynamic discussion. You can exaggerate ("Okay, but what happens if we do nothing") or support a dissenting opinion ("He has a point though, right?"). Jump back in when everything has been said and people are starting to repeat things. The nice thing is that at this point, meeting participants will automatically look towards the chairperson again. Taking back control may be easier than you'd expect.

How can I be less bossy and still be effective?

A senior project leader asks:

I am used to a very hierarchical culture. This year, I started working at a high-tech firm in the Netherlands, a country famous for its "flat" organization culture. Here, people tell me I'm too bossy. Because of my behavior, my team has apparently become too passive and only does what I tell them to do. I do feel as if I am carrying a big burden by myself, without the team's help. How can I change that?

The communication trainer answers:

In the Netherlands, "authoritarian" is practically a dirty word. Here, more than anywhere else in the world, organizations are flat, meetings are numerous, and people will eagerly give you their solicited or unsolicited opinion. The downside is that the decision-making process can be time-consuming and consensus building can produce bad solutions, or no solution at all. The resulting style of leadership and collaboration also has a number of important advantages.

The Netherlands is largely made up of small tracts of land that were reclaimed from the sea and are protected by shared embankments, known as dikes. If your neighbor wasn't taking care of his piece of dike, your life would literally be in danger. Collaboration was an absolute necessity. Personal disputes were considered less important, and people tried to find solutions that were acceptable to everyone. In terms of leadership style, this means you don't make all the decisions yourself. Instead, you make the most of the smart people you worked so hard to recruit. How should you do this?

As an authoritarian manager, you come up with solutions and want everyone to implement them according to your wish. This also means that you need find all the solutions and convince your troops. That may work if you are the wisest, most brilliant person around. But generally you won't be, and you will need to surround yourself with your team's support and input to implement your ideas. Their help will be nipped in the bud if you immediately formulate and impose a solution.

Instead, start by describing the problem as you see it, then move on to the goal you want to achieve, and only then discuss possible solutions. Make sure everyone is involved in the discussion to get all viewpoints out on the table. Once you have all the important pieces of the puzzle, the solution that makes the most sense will present itself. Decide together if you can, call on your authority if you need to.

A workable approach may be to involve your team in the thought process but take the ultimate decision yourself—but only if you need to. By delegating the responsibility for decisions to the lowest possible level, you will maximize efforts and take full advantage of your organization's talent.

> **Decide together if you can, call on your authority if you need to.**

Be selective about the people you consult with. Don't limit yourself to people who will always agree with you, otherwise you won't learn anything. Don't shy away from outspoken criticism. Keep in mind that it's in the company's interest to choose the best solution; this will help you get over your own preferences and any personal feelings.

In summary, it all comes down to the following:

1) Prioritize the company's interests over your personal preferences.

2) Before you make a decision, consult all the major stakeholders, not just "yes men."

3) Delegate the responsibility for decisions to the lowest possible level in the organization.

4) Develop a culture that allows everyone to honestly speak their minds.

How do I motivate my team to develop and perform better?

A team manager says:

I recently started working as a team manager, and I'm facing a team that's afraid of making mistakes and therefore learns very little. I've noticed that people avoid challenging projects because they're worried they may not succeed right away. I've also noticed that they don't give each other enough feedback. I would like to improve this culture and create an environment that promotes innovation, support for mutual development, and ultimately improved performance.

The communication trainer responds:

In sports, we often see coaches looking back at the achievement of their talent with disappointment. Talented athletes who had everything going for them but never quite achieve the success that seemed within their reach. How does that happen?

According to Carol Dweck of Stanford University, this has to do with hidden shortcomings, meaning the wrong mindset. To describe our attitude towards our capabilities, she distinguishes between a fixed mindset and a growth mindset. Athletes with a fixed mindset assume that they have a certain amount of talent that they need to fully utilize and show to the world. Someone with a growth mindset assumes they are on a path of development that allows them to improve their level step by step thanks to effort, feedback on their performance, and the lessons learned from their own mistakes. Research among top athletes shows that in the long run, those with a growth mindset perform better and can develop into elite athletes.

Scientific research has also shown that our brains can grow. The term neuroplasticity refers to our brain's capacity to grow and reorganize. This happens when we make a concerted, long-term effort, for example to learn something new. This is true for both physical and mental capacities. And that's good news, as long as we're willing to make an effort and aren't afraid of making mistakes.

> **We need to be willing to make an effort and to make mistakes.**

How do people end up with a fixed mindset or a growth mindset? This has to do partly with our own predisposition and partly with how we've been raised. If you were given a lot of compliments and assurances about your intelligence and great achievements, you will develop the drive to excel. But since mistakes damage your self image, you have to avoid them. You also have to avoid hard work, because people who work hard may not be so talented after all. When confronted by challenges, athletes with a fixed mindset would rather drop out and focus on what they're good at. Blaming others for their own failures or circumstances, hiding their mistakes—that's all part of it.

With a growth mindset however, it's up to you to put in the effort to develop yourself and improve your capabilities and performance in the long term. Challenges are opportunities to learn. After your initial disappointment, you quickly change gears to analyze the situation and your own actions. This neutral attitude towards what went wrong quickly provides valuable information on how to do better next time. This results in a positive growth curve.

As a manager, do you focus primarily on the team's star players? Are you preoccupied by searching for talent to strengthen the team, instead of spending time on teambuilding and activities focused on development? In that case, don't be surprised if your team quickly reaches its limits. You can encourage a growth mindset by openly discussing the topic with your team. When you compliment or manage people, don't just focus on results but also on the effort they put in, their growth, the way they handled mistakes, and the challenges they've taken on. When you show that you, as a team manager, work hard to improve your skills, learn from your mistakes, and welcome feedback, you will inspire your team to do the same.

How do I talk to someone about their annoying behavior during meetings?

A team manager asks:

I organize weekly meetings that the whole team attends. One of the team members tends to dominate the conversation—he interrupts people and won't stop talking. This is annoying for everyone at the meeting, including me. How can I talk to him so he changes his behavior, without invoking his anger?

The communication trainer answers:

First of all, the team manager is responsible for making sure the team meetings run smoothly. It is your job to talk to the person who is displaying disruptive behavior. This will make meetings more efficient, but there's another reason to step in. If you don't address the issue, people will become more and more annoyed by this employee's inappropriate behavior. With each new item on the agenda, his coworkers will worry that he'll start talking again. As a chairperson, you may even avoid looking his way, hoping that will make him less eager to participate in the conversation. The longer this situation continues, the more annoyed people get. And the more this annoyance builds up, the more difficult it is to effectively address the situation with him.

Addressing someone effectively means that the other person "gets" what you're saying. He needs to be willing to listen to you and your message. You won't achieve that by approaching him when you're clearly annoyed. You will need to approach your employee with a positive, or at least neutral, attitude.

That works best when you ask yourself what the underlying positive reasons may be behind his annoying behavior. An employee who is always talkative and opinionated, may just be really enthusiastic. The point it not whether this is true. The point is that you will be more relaxed and neutral when you think of positive reasons. Your own irritation disappears to the background.

Next, approach your employee and clearly describe what is annoying about his behavior. For example, maybe he immediately shares his opinion every time an option is presented, or perhaps he talks about five times as much as the other attendees. Next, it's also important to detail the negative effects of his behavior. For example, describe that it produces irritation and that you have stopped looking his way when you ask a question, in the hope that he'll stay quiet.

Ask yourself what the underlying positive reasons may be.

Before giving your employee the opportunity to respond to your feedback, give a brief summary and underline his good intentions. "So, you're enthusiastic and like to express your opinion. You do that by responding to everything right away. The effect is that I've stopped listening to you and avoid looking your way." Follow up with: "I don't think that's your intention, but it does happen." Now you pause and give him some space to answer.

Nine out of ten times your employee will agree and ask how to do better. You can agree on an approach together. For example: "Wait until at least two other people speak up first, and then decide if you still need to respond." Plan on evaluating this approach after a while.

To have a successful conversation, you will need to briefly prepare what you're going to say. It's best to have a one-on-one meeting. If you manage to do this in a positive way, you'll see that you can address even the most difficult topics.

What is the best approach to mentoring junior coworkers?

A senior engineer asks:

I often work on projects with younger coworkers, coaching them on how to do their job. I rely on my intuition but don't always know what I should be paying attention to. How can I handle this in a more deliberate—and hopefully more effective—way, especially if I want to express criticism?

The communication trainer answers:

When coaching coworkers, you really have two simultaneous goals. The first is to get the actual work done. The second goal is to help the other person develop. As a senior employee, you will have to repeatedly switch focus during conversations. You will alternately focus on content and on the person. You are creating a triangle between you, your coworker, and the content you are discussing. This results in a "relationship line" and a "content line."

It's best to literally, physically create this triangle. So sit down at a 90 degree angle with your coworker, with the work on the table. This will help to separate the relationship from the content. When you focus your feedback on the content, you literally turn to the work on the table. When you talk to your coworker or want them to respond, focus on them. If you properly shift the focus of your attention, your coworker will follow.

When you criticize the work, literally focus on the work, not on the person. You do this by turning your eyes and shoulders slightly toward the stack of paper on the table. This phenomenon, with the other person shifting focus by following your eyes and shoulder movements, happens almost subconsciously. Move just your eyes, and nothing will happen. Move just your head, and nothing will happen. But when you turn your head and shoulders in a certain direction, the other person will inadvertently shift their attention in the same direction.

This is the same technique you use when you stand next to a projector or flip chart and want people to look at the flip chart. Think of the weatherman on TV, who wants you to look at the weather map. Or think of a negotiating table, which will often be oval ("the Oval Office"). In one of the oldest and most notorious streets of Amsterdam, called the Zeedijk, staring straight at someone would signal that you were either looking for a fight or looking for a date. We're talking "street psychology."

When you sit right across from people and stare at them, they tend to feel pressured, which you want to prevent from happening in a coaching setting. When relationship and content lines cross like that, any work-related criticism tends to be taken personally. The trick is to keep the relationship and content separate. By doing so, you and your coworker will be in the same boat, so to speak. Look at the work together with your coworker and discuss it.

You can be quite critical without the other person tuning out

---

This sense of being "in it together" is something you want to cultivate from the very beginning. Do this by connecting briefly at the beginning of the conversation. This may sound a bit vague, but it just means you ask how the other person is doing and take a moment to really listen. Is everything alright? Next, you move on to the content of the meeting. You could say something like, "Let's take a look at the work, OK?" In other words, start by paying attention to the person. Next, focus on content. Keep alternating.

This way, you stay attuned to the other person during the conversation and create a safe environment. Once you've achieved that, you can be quite critical without it negatively impacting the relationship, and without the other person tuning out.

How do I tell someone the bad news?

A manager asks:

Sometimes I have to share bad news with my employees. For example when we cancel a project or take someone off a project. Or maybe an employee isn't getting the raise they expected, because their performance doesn't justify it. These are not easy conversations to have. I seem to fall into every conceivable pitfall there is. How do I manage these conversations and get my message across without alienating someone?

The communication trainer answers:

When you tell someone something they don't like, they may respond with anger or disappointment. It's unreasonable to expect otherwise. And yet, there are a number of things you can do to make sure your message comes across and the other person leaves the room feeling alright. Let's say you have an employee, John. You need to tell him that his skills don't match the project requirements. You still want him to leave your office feeling okay, as far as possible. What pitfalls do you need to avoid?

Don't beat around the bush. If you start talking about other things first or begin with a long introduction, your employee will get distracted. This makes it even harder for you to address the issue. That's the first pitfall. Another pitfall is to start the meeting by giving compliments to create a friendly atmosphere before you tell the bad news. For example, you might say: "Things have been going really well lately..." This creates confusion for the employee. They can probably tell that you want to say something else, even if you don't. Just be straightforward right from the start.

In the first stage of the conversation, immediately say what you want to say, and be clear. For example: "John, I have to tell you something that you're not going to like. I have decided to take you off the project." Come right out and say it. Then pause. John's natural reaction will be something like, "Huh? But why?" Be prepared to provide one or two solid reasons. Elaborating too much is another pitfall. Don't do it. Briefly say why you've come to your decision. You could say: "I've decided this, because I see you're not achieving enough in your current role." Make sure you have some examples in case you need to illustrate your point. However, this should not become the focus of the conversation.

You are now in the second stage of the conversation, and your goal is to assist the other person in processing the news. There's a good chance they will have an emotional response. Even if John doesn't show his emotions, they're still there. If you don't give them space or don't name them, your employee will leave the room with a sense of disappointment or anger. It doesn't really matter if this anger is directed at you or not. As a manager, make yourself the target for a moment so they can release their emotions. Let them blow off some steam.

Make yourself the target for a moment.

Do this by passing the ball to your employee right after you deliver your message. Say something along the lines of: "I can imagine you're upset about it. How are you experiencing this?" Now your job is to give them some space. Do this by pausing and leaning back a little—break off contact for a moment and relax. This will get them to respond. Continue doing this for however long it takes. Ask follow-up questions now and then, and show your empathy ("right, that's disappointing..."). Let them take the lead and follow. Do they still not believe you? Repeat step 1 by briefly stating your message again, for example, "I've made my decision." This usually doesn't take long, but it can feel like hours. After all, it's uncomfortable to have someone else feeling uncomfortable. You'll simply have to get through this to get to the next stage.

In the third stage of the conversation, you work together to find out what arrangements need to be made and what the next steps are. If you take the time to complete all these steps and allow the other person to work through their emotions, you will get back to "work mode" soon enough.

How do I recognize the three signs of a looming burnout?

A manager asks:

I've seen people in our organization, and even in my own team, take time off from work because of stress and burnout, and this seems to happen on a regular basis. That's difficult for those employees, first of all. But it also promptly creates a difficult situation for our team and our company. The problem is that I don't usually see it coming, and by the time it happens, it's too late. What are some practical tips for recognizing a looming burnout, so I can take action before it's too late?

The communication trainer answers:

Stress is a common human phenomenon that comes out in different ways and is not always easy to spot. Many people don't realize their stress is getting out of control. And even if they do, they may be reluctant to talk about it. On the outside, everything seems fine, but on the inside, a storm is brewing. So is there a simple way to predict where things will go wrong?

There are three conditions to look out for, and when they occur simultaneously, you absolutely have to intervene:

1) too much work,

2) conflict, and

3) not enough sleep. If only two of these conditions are present, people can generally keep going for a while. If all three are present, the situation is unsustainable and people will crash.

The problem is that these three conditions often go hand in hand. For example, if you have too much work and that continues for too long, a conflict can arise all of a sudden. At work, you may not be able to meet all your obligations because of the workload. Or the quality of your work suffers, causing friction. Or maybe you get into a conflict at home, because you spend your evenings working or sitting on the couch like a zombie. That will add to your stress. As long as you sleep well, you can recover every day and keep going. But once you start losing sleep and can no longer recharge, things quickly go downhill.

Let's say an employee has too much work. If he's part of a good team and can let go of his work (so he can sleep well), the risk of a burnout is not that big. Let's say he also has a conflict. Now things are getting harder. Then he starts losing sleep. At this point, his stress starts to spiral out of control. When someone doesn't get enough sleep, they are bound to become less effective, make the wrong decisions, and create more work, which leads to conflict and so forth, and so on.

With too much work, conflicts, and a lack of sleep, you will burn out.

Fortunately, you can find out if these three conditions are present. Analyze the workload by listing all the tasks someone is responsible for. For each task, indicate how much time it would take to do the job well, and calculate the total. If the calculated number of hours doesn't fit into a normal workweek, the workload is too big. Find out if there are any conflicts. Conflicts at work are usually no secret. You can ask about any conflicts at home: "I know you're under a lot of pressure. How are things at home? How is your family responding to this?" If your colleague starts to mumble something, things probably aren't going that well. Also ask whether they're still able to get enough sleep. Pay attention and ask follow-up questions if you need to. You really want to find out how they're doing.

If you come to the conclusion that all three conditions are present, the situation needs to be changed urgently. Reduce the work to a realistic number of hours. Call in support to deal with the conflict at work. If you take these steps, you likely prevent things from getting much worse.

How do I develop a training portfolio that meets our needs?

A business unit manager asks:

I was asked to develop a training portfolio for our organization. My goal is to structure the multitude of individual trainings we currently have. I want to get rid of any deadwood. However, it's not clear what the preferred paths for development are, and I suspect not all developmental needs are being met. How do I apply some common sense structure and logic to this situation?

The communication trainer answers:

Building an adequate portfolio requires a number of ingredients or steps:

1) Define the preferred development paths.

2) Diagnose critical situations in which people are ineffective or completely drop the ball.

3) Compile your development offering. Create an overview of all the courses etc. that are currently being offered and come up with a plan to handle any gaps in your offering. Jot down the most important issues on a sheet of paper first. Then later on, you can drill down further.

Start by thinking about the various development paths for your employees. It's important for these to be in line with the organization's needs, now and in the near future. For example, there is no point in having people develop in a direction that your company does not want to pursue. Say you want your architects to broaden their skills, so the company can broaden its scope. Or you want a group of project leaders to develop commercially. This way, you identify a manageable number of development paths focused on the organization's growth.

Next, decide which situations are critical to an employee's success on each path. A critical situation may be the moment an architect wants to present a diverse group of stakeholders with a proposal. Or maybe a project leader needs to turn a budget overrun into a commercial opportunity with a customer.

Step by step, you will achieve a better understanding of the essential competencies and skills an employee needs to master along each development path. You can do that for each level, e.g. junior, mid-level, and senior. Next, you inventory what type of development is necessary to succeed in critical situations. This means that you will need to determine exactly what employees are doing wrong in those situations. If it concerns future scenarios (e.g. project leaders need to learn sales skills to fulfill new job requirements), you will have to figure out what they are bound to do wrong if they try their best and use common sense.

Let's get rid of any deadwood in our training portfolio

For example, the average project leader from your group may try to "sell" products or services and forget to properly research the customer's real problem. So rather than asking questions, he is focused on messaging. The customer may feel misunderstood and reject the sales pitch. The sales don't increase, and the project leader loses any motivation to become more commercial. So what the project leader needs to learn is to get the real problem of the customer out on the table. This process of analyzing will clarify your employee's learning needs. In the case of the project leader, he needs to stop selling and start asking questions and analyzing what the customer needs.

Once you have inventoried this for the most critical situations, you will have a clear understanding of the existing developmental needs. Select the parts of your offering that align with these needs. Analyze what you can get rid off and which educational needs are not being met yet. Base your plan of action on this. This could include anything from courses to peer-to-peer coaching, supervision, computer-based training, mentoring, and switching workplaces. By basing your analysis on the real-world situation, you can effectively distinguish "need to know" from "nice to know." Focus on the "need to know."

How can I handle the organization's rapid growth?

A manager asks:

As the amount of work is quickly increasing, we need more people. We're concerned that we cannot provide the necessary guidance for a large number of new employees. Too many employees without an adequate onboarding process is bound to create problems. What can I do to make sure new employees quickly settle in and seniors don't get overburdened?

The communication trainer answers:

More work means you will need more people to perform the work. Not providing new employees with the necessary guidance could cause more problems than it would solve. New employees will start working with no clear direction, only slowly becoming more productive, if at all. The knowledge and experience that is already present in the organization are underutilized, and needless mistakes are made. The team's senior employees also need to invest considerable time and effort into coaching new hires, which keeps them from doing their own work. Everything comes to a standstill.

What can you do to effectively manage rapid growth? Start with a clear goal that everyone, including new hires, experiences as urgent. Not only will this clearly spell out what needs work, it gets everyone to hone in on the same thing. This is helpful. It's also important to get new employees to work productively as soon as possible. This gives them self confidence and makes them feel part of the organization.

Moreover, a clear structure is necessary to move new employees through an onboarding process. That structure will contain three elements: technical expertise, the company's "way of working" (business processes), and getting to know the relevant people. Recent graduates will have a grasp of domain knowledge, mechanics, optics, or certain IT techniques. But they will need to acquire the specific proprietary knowledge and experience related to products.

They will also need to learn how work gets done, which processes are used, and what the (technical) design rules based on years of experience are. A list of contacts will need to be created, listing people who can contribute to substantive development or offer insight into the organization. A mentor can serve as an anchor within the new organization. He or she can be the point of contact who helps guide the new hire through the onboarding program.

Part of this onboarding program will be generic and focused on company-wide issues, such as understanding the company's business segment, its corporate culture, or any supporting systems. However, for the most part the program will be job specific and vary by subject area, role, and required competencies. That's why it is useful to develop specific curricula that can quickly be completed by new hires in a given role.

Talent doesn't want to waste time

How easy it is to develop such a program depends in large part on how many trainings and "way-of-working" descriptions already exist. It takes time to prepare documents and develop trainings. In this case, "something is better than nothing." Any help is welcome for new employees who want to get to work faster and better. This investment will certainly pay off by quickly bringing employees up to speed and preventing needless mistakes. Having a clear structure for onboarding new employees also gives your company an edge in the competition for talent. After all, they want to hit the ground running and steer their own development without wasting time.

How do my team and I come to terms with the death of our colleague?

A team leader asks:

Someone in our team recently passed away. We've always been close as a team, and this is a big blow for us. Reorganizing the work is something we can deal with, but what about the emotional process, not just on an individual level, but as a team? How should we handle this?

The communication trainer answers:

Grief is not something you "handle." You allow it. What you can do as a manager, is to make room for it. Everyone deals with feelings of loss differently, so there are no hard and fast rules on the best way to do this. But there are still meaningful ways to address grief. When you're in mourning, you go through a whirlwind of emotions. Anything can come up, from sadness and anger to self-reproach and depressive feelings. Eventually you will have to find a way, both as an individual and a team, to move on with your life and at the same time keep the memory of the deceased person alive.

But there are a number of pitfalls on the way before getting to that point. You can run from your feelings, for example by working harder or looking for other distractions. This can work, at least for a while. You go on with your life. But when feelings aren't processed, the pain that comes with loss has nowhere to go. This can make memories painful to bear. When you push away difficult feelings, you also close yourself off to all kinds of positive feelings. Or the opposite can happen, and you get lost in all the emotions you're experiencing. This can make it hard to function normally. That can be a problem too, since it can affect your work in the long term.

The question is how you can work with your feelings and at the same time find a way to move forward as a team and as a human being. The rule that applies here is "How do you eat an elephant? One bite at a time." This involves the conscious creation of meaningful moments. As a manager, you can support this. Rituals can serve as a way to mark the loss. It's important to do this together and for everyone to have a chance to express their feelings. This could take many forms, and which one you chose it less important.

For example, your get together as a group to collect pictures, share memories, tell a story, take a walk in memory of your colleague, and so on. Reorganizing the work is another key moment that you can discuss together.

Grief is not something you "handle." You allow it.

On a personal level, consider creating your own rituals or space to stop and think about the person you've lost and the feelings you have. Set aside some time to do this. This can provide an outlet. For example, chose an evening to be on your own once a week, turn off your phone, and take some time. This way, you take some pressure off.

The challenge is to let your feelings emerge without holding on to them or getting lost in them. As a team and as an individual you will have to find this balance.

Consultative sales

How can I get better at selling my service?

An engineer asks:

As a testing expert, I often have to work hard to convince project leaders that a particular test is absolutely necessary. I find it hard to convince them, especially if a project leader is acting dismissively. How can I get better at selling my service?

The communication trainer answers:

Even if "sales" isn't part of your job description, you may very well need to "sell" your services to internal customers. Maybe they can outsource your service, or maybe they simply don't need to buy your service or product for example. For many technical people, "sales" is a dirty word. Peddling nonsense or cheating customers is certainly something you should avoid if you want to get a good night's sleep. But selling can also be a way to help a customer solve a problem. How should you do this?

The sales process consists of four steps:

1) Attention

2) Inventory

3) Presentation

4) Closure

In the first step, you approach the customer to let them know who you are and what you and your department have to offer. At this point, it's important to get the customer's attention. So in your introduction, mention something you know or think will be useful for the customer. For example: "We are developing a new test that can detect errors in the XYZ process early on."

Now that you have their attention, you have more or less "earned" the right to ask them questions. You may be tempted to start telling the customer how great your service or product is. That would be a mistake, since you don't yet know if the customer really needs your product or service at this point in time. You are flying blind.

The next step is to inventory the customer's problem or need. What is a current concern they could use your help with? When you ask questions, you want to find out two things: what the customer's problem is and what the customer needs. Do this by asking open-ended questions: "What is the problem?" "What issues are you running into?" "What goes wrong?" "What negative effects does this have?" This way, you can inventory the customer's pain points, and that provides the motivation to take action. Also ask which goals the customer wants to achieve. This builds up a so-called creative tension between the here-and-now ("ouch") and where the customer wants to go ("smiling faces and good cheer"). You add value because you can help the customer take that step.

> **It's only natural that people will ask some critical questions.**

Once the problem is out on the table, you proceed to step 3, the presentation of your solution. Don't be afraid to make a promise. After all, you stand behind your product. If it's too early to guarantee a certain result, name the degree of uncertainty. Either way, take a stance. For example, you can say: "I still see a 10% risk that we may not be able to achieve the desired result. But I would like to go for it and limit the risk as much as possible." Waiting until you are 100% sure of something is science. Making a choice and taking a stance in uncertain circumstances demonstrates leadership. And this will often be necessary.

Don't expect to be applauded right away when you present your solution. It's only natural that people will ask some critical questions. Try to find out what their concerns are. Take a step back to the "inventory phase," and ask open-ended questions until all the customer's concerns are clear. In other words, alternate between messaging (phase 3) and asking questions (phase 2). You will gain your customer's trust once your proposal fits their needs, and they will want to come to an agreement and close the deal (phase 4). You have actually helped your customer!

How do I call a potential customer?

A project leader asks:

My manager wants me to network more. This would involve calling potential customers that I've met at meetings or trade fairs and that I barely know. How do I handle these calls?

The communication trainer answers:

"Cold calling" is something most engineers would rather avoid. That's understandable, because a lot of people feel uneasy about "selling" themselves. So if you are reluctant, you're not the only one. Nevertheless, it's a skill you can learn and may even come to enjoy.

Cold calling means you make a phone call to a potential customer you've never met or have only talked to briefly. A contact is "warmer" if you already know the other person a bit better than that. The better you know someone, the easier it is to pick up the phone and call them. A warm contact is already part of your network, while a cold contact is not.

Networking refers to building relationships that can be mutually beneficial, and customer acquisition refers to actually bringing in work. Developing contacts and discussing what you can or cannot do for potential customers, is a step-by-step process. It starts with sitting down together or at least getting in touch with each other. How do you get there?

First off, have a realistic goal when you make a call. It takes more than one phone call to get a big project. So what is a realistic goal—what is possible? Coming in for a talk is a realistic goal, for example. Or maybe the person you're calling can refer you to someone else, so you can talk to them. Before you make the call, clearly establish a goal for yourself, and make sure your request is as small as possible. So small, that the other person can't really say no.

This has another positive effect: When someone says "yes" once, the brain automatically tries to justify this "yes." In other words: From that moment on, the other person will come up with reasons for why it's a good thing he or she said "yes." The next "yes" will be easier to get.

So what can you say when you call? If you call someone you don't know, introduce yourself. Briefly describe what you do, who your customers are, and what the benefits are: "Hi, I'm Pete from company X, and I specialize in these solutions for medium-size companies. The solution I offer produces a cost savings of twenty percent and up, on average."

Ask for little, and keep it light.

If you've already talked to someone once, for example at a trade fair, mention that. In that case, you don't have to introduce yourself. Once you've established that initial contact, move on to the next step: Get straight to the point. Just tell them what you want or what you want to ask. For example: "I'm working a new concept to tackle this or that type of problem. I would love to hear your thoughts." Or: "I would like to come and talk to you about a specific solution to a problem you may be dealing with as well."

It's important to keep the conversation light. If a moment presents itself to laugh together, that would be great. Don't be afraid to sell yourself, and realize that you really have something to offer.

And last but not least: Be straightforward. Don't tell someone they really should talk to you because they would be better of. No one wants to hear that. Just say you would like to talk to them, because you think you have something to offer. Be prepared, ask for little, and keep it light.

How do I persuade potential customers to meet with me, even if I don't know them?

A team manager asks:

Management has told us that we need to work on expanding our customer base. This means I will have to search for customers I have never met. How should I handle this? I wouldn't consider cold calling one of my favorite pastimes.

The communication trainer answers:

Cold calling means you call a potential customer without any prior contact and ask to meet with them. Most professionals will break out in a cold sweat at the thought of it. The resistance to cold calling is largely due to a lack of knowledge about the right approach. How do you make a cold-call in a pleasant and effective way?

The key is to limit yourself to setting up an appointment for a conversation. Do not call expecting to make a sale—that's certainly not the goal. Your goal is to set up a meeting to discuss what you have to offer and what may be interesting for the customer.

Cold calling involves two important steps: preparation and the actual conversation. During preparation, you start by searching for companies that fit the customer profile and would benefit from your services. Do this by visiting the websites of trade associations, the Chamber of Commerce, network organizations, etc. Find out what kind of work they do, and also pay attention to the organization's size, revenue, and even location—after all, if they become a customer, you should be able to easily reach them.

Next, find out the name of the decision maker you could actually do business with. For small and medium-sized companies this is usually the CEO, and for large enterprises this might also be the head of a department. Don't aim too low. A name and direct phone number would be great, but otherwise just use the name so your call can be transferred.

To prepare, think about what you are going to say to get an appointment. Your story will contain a number of elements: a greeting, a pitch about what you have to offer, a few words about what you want (to come in for a talk), and a specific proposal for a time to meet. Keep it short and to the point. Thirty seconds should be enough. For example: "Hi, do you have a moment? I am so-and-so from company XYZ. We recently developed a new product that does A and B. Our goal is to expand the customer base for our product. I would like to tell you more. Would you have time to talk next Tuesday or Thursday morning?" That's it.

The trick is knowing how to handle responses.

The trick is knowing how to handle responses. When you ask if they have a moment and the answer is "no," your reply is: "When would be a good time?" If the other person says: "Never," respect the answer and end the call. If he suggests a different time, set up a time for a call. Say the answer is: "If you can keep it short," you reply: "Yes, I can keep it short..." and you continue your story. At the end of your story, the other person will often ask: "Alright, but what exactly does your company do?" This gives you the opportunity to provide more information. Go ahead and do that, but keep it short, and end by saying: "I would like to continue our conversation. Would next Tuesday or Thursday morning work for you?" Always end with a specific proposal. You've done well if you can come in for a meeting.

No harm in practicing ahead of time with a co-worker. And after an afternoon of making calls, you can review how often you talked to the right person and managed to get an appointment. Best of luck!

How do I put pressure on a negotiation without shutting the door?

A buyer asks:

During negotiations with suppliers, I'm often reluctant to put pressure on the other party. I worry about damaging the relationship, along with the project. I would like to work things out with the supplier, partly because it is hard to find alternatives. How can I increase the pressure without shutting the door on negotiations?

The communication trainer answers:

Needless to say, the strongest position to negotiate from is having a clear BATNA or "best alternative to a negotiated agreement." Being able to easily switch to a different supplier, for example, makes it easier to put pressure on the other party. Yet, in practice, switching suppliers is often cumbersome or something you would rather avoid.

What can you do to reach an agreement and apply some pressure without putting the relationship as risk? During negotiations, it is crucial to find out what the other party's interests are. You may have different standpoints; for example, you may not want to pay what the supplier wants to charge. Or maybe the supplier will be developing a technique that he wants to use for other customers too, and you want to prevent that. By focusing on these standpoints and the differences between them, positions quickly solidify. So find out what the other party's underlying interests are.

The need for a robust profit margin, having enough latitude for developing competencies, protecting your own business, increasing market share, or being able to follow your own operating procedures: these are all examples of interests that may be at play in the background. Once you know what's at stake for each party, it opens up space and creativity for finding solutions both parties can benefit from. If this is not the case, techniques for applying pressure during negotiations are geared towards achieving a breakthrough and renewed creativity.

How do you do that? The first technique is "rational influencing." You do this in three steps.

1) Indicate what the negative effects would be if, for example, certain specifications aren't met. You might say: "When the proposed specifications aren't met, chances are that this or that will go wrong with the product. We can't take that risk."

2) Formulate a (common) goal, e.g. "We want the product to work and fulfill the customer's requirements."

3) Repeat your proposal and explain why you would like to stick to it: "That's why I propose we..."

Increased pressure can create another breakthrough.

If you have examined all the options together and the answer is still "no," you can apply pressure by making a personal request. Call on the loyalty, goodwill, and relationship you have with your negotiating partner. Be reasonable. You could say something along the lines of: "I would like to continue working together and make this product a success. But I can't help but notice that our negotiations have been going on for two months, and I am losing confidence in our ability to work things out. It would be too bad if we can't. What possibilities do you see at this point to come to a solution?" Here's another example: "We can bicker over these negotiations. But I would rather be open and transparent and would ask you to do the same. What I'd like to do is decide on a price that is reasonable for both of us. I realize this may be difficult for you, but what do you think?" The secret is to really let go, relax, and consciously stop talking after you say this. The other person will need a moment to think it over.

You can apply even more pressure by addressing the dilemma. You will need some nerve to do this. Say: "I would really like to, but it seems we're not finding a solution this way." Then let go, relax, and wait. If someone doesn't want to go along, the partnership apparently does not matter to him. It's better to be clear about that. Making the dilemma explicit will usually create an opening or breakthrough.

Self-management

How do I make sure I'm doing the important things at work?

A team manager asks:

I've noticed that at night I often worry about important things related to my work. It's a problem: I don't sleep well, and it makes me feels stressed. How can I keep myself from lying awake at night worrying about work, and ensure that I give the most important thing the attention they deserve during the day?

The communication trainer answers:

Your daily routine keeps you focused on what *has* to happen now – or seems like it does. These are often urgent matters, which demand immediate attention and dial up your sense of pressure. If you aren't careful, you'll give in to them easily, and your working memory will fill up with all kinds of mundane tasks.

The result is that the larger issues that are more hidden from view fade into the back ground. A work process isn't functioning optimally; you have a good idea for a new product that you haven't gotten around to yet; you're on uncomfortable terms with a colleague; a conflict is still festering under the surface; you wonder if you're in the right place. If you can spend time on these things during the day, you can keep them from growing until they disrupt your nights. But how exactly do you do that?

All kinds of signals give you information about what's going on behind the scenes and needs attention. These signals are very subtle for a long time, then suddenly start to blare. If you don't realize your car is out of oil until you're stranded on the side of the road, you're too late. The damage is massive, and it will cost you much more energy to fix the situation. The earlier you realize that something's wrong, the better. But that only works if you pick up on the subtle signals (such as the small dashboard light that's come on) and then actually act on them.

> **Stepping back and reflecting doesn't cost you a thing.**

To pick up on these subtle signals early, you'll first have to recognize them and learn to assess their significance. It's easiest to notice them when your mind is quiet. So, throughout the day, create moments of distance and rest. For example, take a short break every day and consciously lean back in your chair, take a walk alone, sit somewhere quiet and drink a really good cup of coffee. Relax, and then ask yourself these questions: 'What's going well?', 'What's not going well?' and 'How can I make things better?'. Questions like 'Who would it be good to get back in touch with?' can also spark some good ideas.

If you manage others, then walk around and check in with team member every day. Be sure to first take time to relax. That activates the best parts of your brain: your creativity, your emotional intelligence, your problem-solving ability, your attention to detail. After making the rounds, sit down and ask yourself whether you need to provide guidance anywhere.

Stepping back and reflecting on your team's work and on yourself doesn't cost you a thing, and it helps you make sure you're busy doing the right things. What's more, it helps you maintain a healthy work-life balance.

Make sure you have a pen and paper with you during the moments when you take a step back. That way you can always write down any thoughts that come to you. It is work time, after all, and maybe the most important work time of the week.

How can I remain effective in chaotic circumstances?

An engineer asks:

As a project leader, I often work under pressure and am constantly dealing with changes. In these chaotic situations, there are many things happening at the same time, and I lose track, start making errors, and forget details. The problem is that this makes it even harder to keep things under control. How should I handle this?

The communication trainer answers:

It's not so much the situation that's chaotic, you become chaotic. The situation is just the way it is. Because your mind is no longer calm, you lose track and are unable to apply all the project management tools in your arsenal. This problem has a number of root causes. It is based on an important misconception. We don't fully realize that the things change, sometimes rapidly and simultaneously. Change can often make work interesting, except when you try to control everything and become frustrated when that doesn't work.

A number of things are important for effectively managing rapidly changing situations. First, it's important to realize that it's in the nature of things to constantly change. This may sound like a no-brainer, but in reality, it is a difficult concept to truly accept. We simply assume that everything around us, including us, is permanent. But that's not the case.

Everything is changing in front of your eyes. The job you have, the people you work with, even the coffee mug you're drinking from. That mug may seem like a fixed thing, but it owes its existence to various causes and conditions that come together. When you look more closely, what you call a "coffee mug," consists of various molecules that in turn are made up of all kinds of tiny particles moving in a particular arrangement. And all of this is subject to change. If you drop the mug and break it, would you still call it a mug?

Things change. Seeing them as permanent is unrealistic and results in surprises and stress. Once you really absorb the fact that everything changes, you will be less stressed out about it and relax. The ability to stay relaxed is crucial.

Here, it is useful to distinguish between two aspects of your consciousness: thinking consciousness and consciousness itself.

> *In mere seconds, your mind has made up a whole story.*

Thinking consciousness would like everything to stay the same. When that's not the case, it evokes all kinds of emotions and thoughts. Your expectations aren't met, you have certain thoughts about that, which in turn gives rise to other thoughts and emotions, etc. In mere seconds, your mind has made up a whole story. And your thoughts just keep going. In the meanwhile, you lose track of what's actually happening, and you act based on the emotions, thoughts, and ideas you have about the situation. The result? More turmoil and fewer results.

If you manage to stay relaxed instead, you will get less caught up in your thinking and just stay present with what's happening. When you succeed doing this, a number of positive things happen. Your perception isn't reduced, you can keep track of what's going on, and you are able to see the facts rather than your interpretation of the facts. Your brain remains creative, so you can think of a solution for whatever comes up. Your emotional intelligence continues to function, which means you can stay empathetic, are able to listen, and don't get snappy or stressed in your communication. In other words, you have everything you need to be a leader in a difficult situation. Being you at your best.

A practical advice: focus about 25% of your mind on the task itself, 25% on being aware that you are still focused on the task and not distracted, and the other 50% on staying relaxed and open. Everything will go more smoothly when you manage to do that.

How do I become more patient?

A development engineer asks:

During my most recent performance review, my manager mentioned that I'm often too impatient. I do have a tendency to take over work from coworkers when I think they're not doing a great job or they're too slow. A few of my colleagues are annoyed by that. On top of that, I sometimes miss my own deadlines because of all the extra work, which is stressful. How do I remain patient?

The communication trainer answers:

"Count to ten" sounds easy, but in real life it can be a hard thing to do. Why is that? In essence, patience is the ability to not immediately react when something evokes a reaction. You're irritated, you're in a hurry—whatever you do in that state of mind will generally fail to deliver optimal results. Patience, even if it is just a few seconds, is the only medicine. How does it work?

Let's say you see a colleague working on a project. You're watching him as he tells you what he's working on. At some point, this triggers a thought, for example: "He's not getting anywhere." That thought leads to the next thought: "I have to do something, or it's going to be a mess." The initial thought provokes another thought, and so on and so forth.

All those thoughts and emotions quickly create a story about what's going on. We superimpose or "project" this story onto what's really taking place. All in a matter of seconds. The image thus created can be far removed from reality. Next, we act based on our version of reality. That version may be totally wrong and exists only in our minds. If we respond to a made-up reality, we risk making some really big mistakes.

Impatience means you want things to be different in that very moment. It's like you're fighting with reality, which is always a losing proposition and creates stress. Once you become impatient, you no longer engage with the situation with your heart and mind.

The key to patience is learning how to relax and accept what's happening in the moment. Ignore your train of thought and focus on the factual situation.

That doesn't mean that you give up and let things go. It means you keep a cool head and an open heart, even if things aren't going the way you want them to.

If you stay calm, you can respond quickly or you decide to wait for a moment. Either way, you will be better at seeing the reality of the situation, instead of a version of reality distorted by your haste and irritation. And whenever you're better at seeing the facts for what they are, your actions will be more effective.

Ignore your train of thought.

By the way, in some situations feelings of irritation or anger are right on point. For example, if someone tries to steal your wallet, the situation may call for a strong response. If you manage to act on that and still keep a clear head and an open heart, you can keep observing what you're doing, and you may even have a better aim.

Back to that coworker who is working on a task you delegated. It may be best not to intervene right away but give him an opportunity to take responsibility. Sometimes it's just a matter of a few seconds. Wait for a moment, and give the other person some space, instead of taking the initiative or stepping in.

Yes, you will have to learn to count to ten if you have a tendency to get impatient. Relax during stressful moments, and stop your train of thought before it gets going. Simply ignore that whole series of thoughts and stay focused on what's really happening. This way, you create space to act, stay focused on the factual situation, and automatically become more effective. And that's a win-win situation for you and your colleague.

How do I come across as more senior?

An engineer asks:

I'm a senior-level employee. Everyone who knows me is aware of that. Now that I'm in my forties, my age is also starting to reflect my status. And yet I often feel brushed aside as if I'm a junior employee. Last week for example, I attended a meeting with a customer. He grabbed me by the shoulder in front of all the other attendees and said something like: "Yes, you young folks can learn a thing or two here." What can I do about this?

The communication trainer answers:

Seniority comes along with more insight and experience. And you want to use all that insight and experience in your work. It can be pretty frustrating if you can't do that because you don't come across as senior.

The general impression you make has a lot to do with your attitude, how you act, and how you speak. There are a number of things that can make you appear more junior than you are. Lets start with clothing. Wearing a T-shirt and jeans can make you look like a student, especially if your colleagues and customers are all wearing dress shirts and formal trousers. It's best to be proactive and make sure to adapt the way you dress to the people you work with. Don't overdress—you don't want to be the only one wearing a three-piece suit while everyone else is wearing dress shirts. When adapting your style, remember you are not your clothing. It's just to fit in with the overall culture.

Next consider the way you use your voice in terms of tone, volume, rhythm, and intonation. A common error is to end all your sentences with a rising pitch. It's as if you're asking a question, even when you're making a statement. Listeners will think you're not very convincing, even though they don't know exactly why. So end your sentence with a period and a downward pitch. The more calm you radiate, the more senior you will appear.

The same is true for body language. Think of a playful little puppy that jumps up and down and moves around excitedly. It's typical "youthfull" behavior. An older animal will exude more calmness and authority. That doesn't mean the older animal is no longer able or allowed to play. On the contrary. The best professionals combine inner peace with a lively spirit. This will manifest itself for example in the ease and calmness in your posture while you sit or stand.

Laughing while there's no reason to laugh is distracting. But should you look grouchy? That would be a mistake too. The key is to be aware of what you're doing in the moment you're doing it. For example, you can laugh spontaneously or accentuate that you're relaxed, because as a manager, you want to reassure your team in a difficult situation.

> **Be aware of what you're doing the moment you're doing it.**

Do you have some underlying need to be liked, or are you looking for affirmation? People will often be able to tell. This "needing something from another person" automatically makes you less independent and more junior. The parent-child dynamics immediately comes into play. Sometimes you see this in a slightly tilted, upward head posture during stressful moments. This immediately makes someone seem "junior," so if this is you, make sure to keep your head up straight.

Try to trust you have the capabilities to deal with the challenges at hand, and always ask for help if necessary, as all responsible professionals should do.

Sense that you are truly standing on your own two feet, in the knowledge that we all need each other. Be aware of your inner sincerity and, without getting pretentious, express this with your body language, voice, and behavior.

How do I develop more empathy?

A team manager asks:

My manager and coworkers think I need to be more empathetic. They say that when we work together, I tend to see things from my own perspective with little consideration for others. I do recognize that, but how can I change?

The communication trainer answers:

The essence of empathy is the ability to understand things from someone else's position or perspective. Some people do this automatically, but they may also have trouble being confident, knowing what they want, and standing up for themselves. Others have the exact opposite approach. They know what they want and go for it, without thinking about the effects on others.

Developing more empathy starts with the understanding that others have the same basic needs that you have: to be happy and avoid suffering. In work situations, this translates to "How can I be successful?" and "How can I prevent stress?" Let's say you're talking to a colleague from another department. You don't really know her, but you have to work with her. Ask yourself what it means for her to be successful. What is she paid to do, and how does her manager hold her accountable? What would be an unacceptable result for her to report back? The moment you can imagine that, you understand where she's coming from. Being able to imagine what is important for the other person is a matter of logical reasoning.

This results in a more open mindset towards the other person. You won't be so quick to think: "Right, that's one of those... so he'll probably..." Instead, you will be able to observe the facts of a situation with an open mindset. Let's say someone is pressuring you, and you understand why. That doesn't mean you give in right away. On the contrary, you will be more successful at starting a real conversation. You understand what's important to them, but in this case you may simply believe your own interests are more important.

Let's look at another example. You are telling a coworker what you want: "So I believe this is really necessary..." You notice his eyes are glazing over, or maybe he is getting ready to object: "Well, I don't think so. I believe..." This should set off a little "empathy alarm." Instead of repeating your message, shift your attention 180 degrees and focus on understanding his interests instead of your own. Say: "Tell me more…" Forget about your own agenda for a moment. Try to find out what's important for the other person and why he's concerned about what you said.

Instead of repeating your message, shift your attention 180 degrees.

Once you understand his position, you can continue making your point. Your coworker will also be more open to listening to you. Your willingness to understand the other person's point, is not something you can fake, and it will require a real effort.

If you really listen to others, your own opinion may shift a bit too. And even if it doesn't, you will still be better at connecting your arguments to their story and their needs. Either way, you are more likely to find solutions that work for both of you.

How do I keep developing without changing jobs?

An engineer asks:

I have grown quite a bit in my role and have gone through most of the major challenges that come with my job. I don't want to come to a standstill. My ambition is not to change jobs, but I would like to work on my professional and personal development. How should I handle this?

The communication trainer answers:

It's an interesting question: Should you always continue developing yourself? If you feel good and your competencies match your job requirements, you are in the right place. In that case, your managers may not motivate you to keep developing or try a different role—they're happy that everything is running smoothly, without any issues they need to solve. But sometimes, when we get restless and do change jobs, we realize we should have probably moved on sooner. Sound familiar? Stability can lead to stagnation. Staying curious and trying new things can keep you fresh. It can also make you flexible and give you the wide-ranging skills that make you easily employable when circumstances change.

You don't necessarily have to look for a new job to develop yourself. Personal competencies that are instrumental to your professional success can be developed outside the workplace too. For example, you can volunteer, play sports, or take courses. Three critical competencies are assertiveness, the ability to empathize, and creativity. You develop these competencies primarily by exposing yourself to different experiences.

Learning takes place outside your comfort zone

Let's start with assertiveness. Assertiveness is learning how to act forcefully while respecting your own interests and those of others. A lot of people don't act forceful enough, while others are forceful to the point of being rude and steamrolling others. The challenge is to clearly say what you want while staying engaged with the other person. A good martial arts school can teach you how to use just the right amount of force while remaining connected with the other person. Examples of martial arts are aikido, full contact karate, and judo. When you look for a school, make sure the teacher has a good heart. You can recognize this by a sense of mutual respect and safety.

The ability to empathize means that you can listen to others and see things from their perspective. You can develop this quality by helping people who need some extra care. There are all kinds of opportunities to do so: You can shop for groceries for elderly people who live alone, help out in a homeless shelter, or volunteer for a mental health helpline.

You can learn to be more creative and spontaneous by participating in some form of improvisational theater. This teaches you to respond to situations almost without thinking and add your own creative impulse. You learn to deal with every situation spontaneously and to express yourself.

Last but not least is your ability to stay calm and keep track of what's going on in the here and now. You learn this be taking for example a mindfulness course. A mindfulness course will provide tools that you can use in your work too.

Developing your assertiveness, your ability to empathize, and your creativity will not only make you stronger personally, it will also make you better as a professional. Many of the courses mentioned here are fairly inexpensive and can offer a lot of value. No need to become a karate master or an actor. Just focus on participating and learning. The point is to challenge and develop yourself. Learning takes place outside your comfort zone.

How do I retain the skills I have learned?

A team leader asks:

Last month, I participated in a training on conversational skills. It was an excellent training, but I have noticed that the material is already fading from my memory, and I am not happy about that. How can I retain what I have learned?

The communication trainer answers:

We are all in the same boat. We simply forget new skills, forget good intentions, and forget the insights we have gained. This is particularly true when it comes to breaking bad habits. Habits so ingrained, that they have almost become part of who we are. The advantage of habits is that you no longer have to think about them. Over the years, you become so well-trained at something that you can almost do it automatically. A habit is like muscle memory, saving you a lot of energy.

That is exactly why change is so hard and why you will return to old patterns time and time again: You are not paying attention. In order to teach yourself something that is not a habit—as is the case when you learn something new—you have to pay deliberate attention. By being aware of what you are doing when you are doing it, you are creating an opportunity for course correction.

Being and staying aware of your actions can be difficult in situations in which you usually close yourself off, get angry, act without thinking, and basically do the opposite of what you have learned. Your awareness narrows, as it were, and you are left seeing only part of the bigger picture. Suppose, for example, that you tend to give unclear feedback. Maybe providing feedback makes you nervous. You are fully focused on the conversation, and you no longer think of the feedback rules from your training.

If you really want to master a new skill, you need to put effort into reminding yourself to apply that skill. That will make you aware again, so you can correct your behavior. The fact that you are unhappy about forgetting your intentions, is a good start. You are paying attention.

To turn your new behavior into a habit, you will need a few ingredients. First and foremost, you will need the willpower to get started. If you lack the motivation for change, it will be difficult to sustain your efforts. Giving this issue some serious thought is a worthwhile effort. What are you doing it for? Even better, include others in your motivations. You are not just doing what is right for you, you are doing right by others. For example: "I want to manage my workload better, which means limiting my work to eight hours a day and occasionally saying no, so I can spend more time at home with my family." Your motivation can keep you from falling back into old habits.

Your motivation can keep you from falling back into old habits.

This only works if you actually recall what motivates you during difficult moments. Therefore, it is crucial to put in place a system that reminds you of your desired behaviors and the motives behind it. The goal is to create moments of awareness as often as you can—awareness of your actions, their consequences, and the goals you would like to achieve. A nice example is keeping a framed picture of your child in your car, accompanied by the text: "Think of me, daddy." Why keep it there? Every time you see that picture, you are reminded that getting home safely is more important than driving fast.

You can put sticky notes with reminders on your computer, add reminders to your planner, or keep a daily log. Definitely also talk to friends and colleagues, so they can help remind you and provide feedback. In practice, this social aspect has proven to be equally important. In other words, if you want to change, don't go it alone, and involve others in your mission.

How do I show leadership and take a step towards a senior position?

An architect asks:

I am expected to show more leadership and grow into a senior role. The expectation is for me to "really take responsibility, show who I am, and take the lead." Apparently, I'm still not doing that in spite of all the courses I've taken. What do they mean exactly? And what can I do to take this next step?

The communication trainer answers:

Your development from junior level to middle and senior level is gradual and will not happen without any bumps in the road. Generally, going from a junior to mid-level position is easy. You know that you know very little, and you are ready to take on anything. This way it doesn't take long for you to gain the insight and skills that come easy to you.

Once you have a mid-level position, things are usually different. The issues are a bit more complicated, things may not always go smoothly, and you might be less interested in some of the work. Often without realizing it, you cannot yet fully assess the complexity of your task. As the great soccer legend Johan Cruyff used to say: "You'll see it once you understand it." You don't really know your own blind spots. This makes it harder to steer your development.

Those in a senior level position will know what they know and don't know. But how do you get from a mid-level position to a senior position? And what should you do to get there? "Taking more responsibility," "showing who you are," and "taking the lead" are typical senior qualities, but how do you demonstrate these?

What they really mean is that you are willing to take the risk that things could go wrong, with all the damage this could do to you and the project. It's important to realize that the step from mid-level to senior level cannot be made without damage or the risk of damage. After all, you already know how to do the easy work, but now things will get harder.

Maybe this means you will take charge of a project that is essential for the company and attach your name to it. Or maybe you'll have the courage to make a statement that will be the basis for an important decision, even if you're not certain because many fact are still unknown. You stick out your neck and take on a difficult problem. In other words, you step out of your comfort zone.

Why is this so hard? It's all in your head. We all have thoughts that stop us from taking the necessary steps to grow. Fortunately, there is a remedy for any type of thought. A common limiting thought is "What if things go wrong?" This thought will keep you stuck in your current position. Just remember this when you take the plunge: If avoiding failure is your main concern, success is impossible. Simply because you won't even try, and if you do, you are too stressed out to be your best possible self. And you need to be at your best to be successful.

> *If avoiding failure is your main concern, success is impossible.*

The opposite is often true too. In that case you might think "What if I do succeed?" The fear of success can be larger than the small spaces you're used to. The adjustment to a "successful" self image, along with the corresponding visibility and responsibility, can subconsciously stop someone from taking on challenges and even lead to self-sabotage.

How do you handle this? Start by becoming aware when this is happening. Next, force your own hand: Put yourself in a position that leaves you no other option than to move forward. This immediately raises another seniority issue: Make sure you know what support you need to surround yourself with. Is there a mentor who can guide you? Do you have the support you need at home? Can you withdraw from the assignment if you realize that it's impossible (not because you can't do it, but because it is simply not feasible)? If you have the nerve to do this, you are adequately prepared and ready to go.

How do I apologize?

A software designer asks:

I recently had to admit that I made a big mistake while working on a project. This has damaged my relationship with a number of team members. How can I repair the damage?

The communication trainer answers:

We all make mistakes, no matter how good we are at our job. You could say it's impossible to grow without making mistakes. Certain mistakes can have direct negative consequences for your coworkers, and that can be a problem. Enough of a problem that continuing as if "nothing happened" is not an option. Trust has been damaged. And trust is essential: It is the foundation of a good working relationship. The more trust, the smoother things will go.

But we all know it is easier to lose trust than to build it. So when you lose someone's trust, it is very important to try your best to restore it as quickly as possible. You may need to apologize in order to do so.

Offering a sincere apology is not easy. Your pride gets in the way. Many people would rather move to a desert island than apologize. Your ego is not going to like biting the dust, but sometimes an apology is the only way to repair the relationship. Besides, you can clear the air by admitting you regret something. It puts you back on the "right path," as it were.

Although you cannot fake having regrets or offering a sincere apology, you can try your best to do it successfully. A good apology consists of three steps: The first step is to ask the other person exactly what they find fault with. What are you being reproached for specifically? You will need to ask probing questions, even if that feels difficult. Ask questions about the facts and about any feelings that may be present. You may be tempted to start defending yourself. Don't. Just listen and ask questions. In step two, you summarize the criticism and check if there's anything else ("So you're reproaching me for so-and-so. Is that right?"). In step three you sincerely tell them you're sorry. It's important to leave it at that. Be silent and just wait for the other person to respond. It's now up to them.

You regain trust by demonstrating you want to improve things.

A sincere apology will often allow you to move on together. The relationship has been repaired, but the same may not be true for trust. This will require taking real action in line with what you've agreed on. In other words, you will have to demonstrate that you're serious and want to improve things. You will have to show you are taking responsibility and will hold yourself accountable. For example, tell your coworker how you are handling things now, so they can start trusting you again based on facts. It's important to do this at your own initiative, so people don't have to ask you for it and you prevent difficult questions from coming up.

Yes, it would be better to prevent errors from happening in the first place, but that's not always possible. By following the steps above, you have a very good chance of restoring trust.

More information about Jaco and his writings & training

Connect with Jaco Friedrich via LinkedIn or if you have a question please find him via j.c.friedrich@iniziotraining.nl.

More information about the follow up book on "Leadership skills for engineers: 50 more questions" and new columns you find on www.jacofriedrich.nl.

The newest communication and leadership trainings you find at www.hightechintitute.nl.

You help the Leadership Skills to spread if you leave a recommendation.

Thanks!

Disclaimer: Although the author has made every effort to ensure that the information in this book was correct at press time, and while this publication is designed to provide accurate information in regard to the subject matter covered, the author assumes no responsibility for errors, inaccuracies, omissions, or any other inconsistencies herein and hereby disclaims any liability to any party for any loss, damage, or disruption caused by errors or omissions, whether such errors or omissions result from negligence, accident, or any other cause. The author makes no guarantees concerning the level of success you may experience by following the advice and strategies contained in this book, and you accept the risk that results will differ for each individual. The testimonials and examples provided in this book show exceptional results, which may not apply to the average reader, and are not intended to represent or guarantee that you will achieve the same or similar results. Your use of the information in this book is at your own risk. Unless otherwise indicated, all the names, characters, businesses, places, events and incidents in this book are either the product of the author's imagination or used in a fictitious manner. Any resemblance to actual persons, living or dead, or actual events is purely coincidental.